D1506804

Video recorders

Video recorders

Principles and operation

Professor Z. Q. You
Dr T. H. Edgar

Prentice Hall

New York London Toronto Sydney Toyko Singapore

First published 1992 by
Prentice Hall International (UK) Ltd
Campus 400, Maylands Avenue
Hemel Hempstead
Hertfordshire HP2 7EZ
A division of
Simon & Schuster International Group

© Prentice Hall International (UK) Ltd, 1992

Typeset in 10/12 pt Times
by MHL Typesetting Ltd, Coventry

Printed and bound in Great Britain
by BPCC Wheatons Ltd, Exeter

Library of Congress Cataloging-in-Publication Data

You, Z.Q.
    Video recorders: principles and operation / Z.Q. You, T.H.
Edgar.
        p.   cm.
    Includes index.
    ISBN 0-13-945890-5 : $55.00
    1. Video tape recorders and recording.   I. Edgar, T.H.
II. Title.
TK6655.V5Y68   1992
621.388'337--dc20   91-34573
                                                                      CIP

British Library Cataloguing in Publication Data

You, Z.Q.
    Video recorders: Principles and operation.
    I. Title   II. Edgar, T.H.
    621.388001

    ISBN 0-13-945890-5

---

1   2   3   4   5   96   95   94   93   92

Contents

Preface

In just a few years video technology and the reproduction, processing and storage of video signals has been developed to provide users with very high quality video images beyond that initially envisaged by early pioneers. The pace of development and the use of video equipment in the domestic consumer market is increasing due to a myriad of reasons.

Some of the reasons lie with the development of microelectronics and the consequential increase in sophistication and the reduction in volume costs. These factors increase the availability of, and access to, lightweight, low cost video cameras and recorders. There are few homes, colleges or schools without access to a video recorder of some type. Personal ownership of video cameras is now fairly common and they can be seen at almost any holiday resort or celebration venue, recording events which in the past would have been recorded only by the 8 mm movie film enthusiast.

Growth of ownership and use of video recording and reproduction facilities has also led to greater production of commercial material for entertainment viewing and the growth of a market for film and equipment hire and sale. Inevitably, sophisticated equipment of this type will require repair or servicing. This in itself leads to problems for those in the service industry because of the very fast pace of development associated with such a dynamic consumer field.

Service manuals are usually readily available for most video cassette recorders (VCRs) and cameras, and are necessary for the diagnosis of faults or the treatment of service routines, but they rarely describe the principles of operation of the equipment. Textbooks can also be very limited since they often treat video as one chapter in a book about communications or signal processing. The result is superficial with regard to the principles of video recording and reproduction.

This book is aimed at describing the fundamental principles which lie behind the operation of video recorders to aid service personnel and enthusiasts to better understand the operation of these devices. The book will also be of considerable value to undergraduate students pursuing a course in communications in which video principles are a part. Descriptions are given for the operation of VCRs which are designed for various formats, and the commonalities and differences are particularly highlighted. Formats include VHS (video home system), Betamax, U-matic (high and low band) and the 8 mm format. Digital techniques and special function facilities are also included.

It is assumed that the reader has a basic knowledge of electronic devices and their operation, and has the mathematical ability to understand the formulation of modulation methods. The mathematical treatments are available in the appendices.

The reader is led through a basic introduction to audio and video recorders, the principles behind recording and playback using magnetic material for storage, and the various scanning techniques used to record and recover the video and audio signals. Practical requirements for recording a video signal are discussed and the bandwidth limitations are highlighted.

Chapter 1 is a general introduction to video recorders, while Chapter 2 is devoted to the description of the modulation and frequency conversion methods used in storing and retrieving video and audio signals using a magnetic tape. The various servo systems used for control in the VCR are also discussed and the reader is given explanations of tracking requirements to ensure quality picture reproduction.

Video signal processing and the circuits involved in the video channel are discussed extensively in Chapter 3. The operation of a range of circuits used in the video channel are discussed in detail and the principles are used to place their operation in the context of achieving error-free recording and replay. Both the recording and replay paths are discussed. This chapter also deals with the corrections needed for tape imperfections and noise generated in the modulation system. The fundamental principles behind these methods are dealt with in detail.

Chapter 4 discusses the servo control systems and their role in the VCR. Operation of the principal drive servos are described in terms of both their mechanical and electrical features. The discussion also includes the differences in design and operation between U-matic and VHS servo systems.

Both VHS and U-matic mechanical and electrical system control are described in Chapter 5. Operation of the microprocessors and their signalling is discussed in terms of display and system function control.

Chapter 6 is devoted to the treatment and comparison between analogue and digital special function operation. The VCRs which utilize additional heads for still, long play and slow motion replay are described and the principles behind the tracking requirements for these modes are given detailed treatment. Analogue to digital and digital to analogue conversion is also discussed, and the reader is introduced to the sampling and quantization principles needed to understand the digital processing of an analogue video signal.

Principles behind the latest developments in standard and high band 8 mm VCRs are introduced in Chapter 7 and are supported by discussions on improvements in the video channel with the adoption of these new techniques. Methods using frequency modulation (FM) and PCM recording of the audio signal in these VCRs are also outlined. This chapter also includes the principles behind correcting errors in recording and recovering digital audio signals, and the methods used to encode the signals in the PCM audio channel.

Extensive use is made of block and system diagrams for most formats throughout the book to support the understanding of the basic principles embodied in the recording and recovery of video signals. The diagrams also help to place the component functions in a system context.

Finally, although not covered in this book, the reader should be aware of a whole new field emerging in the storage and manipulation of video images. The video disc has been around for a few years but failed to gain any significant market for the replay of conventional linear video. The technology suffered some early problems related to surface coatings; the discs were expensive to produce for low volume sales and contained only approximately 30 minutes of video per side. However, the situation is changing rapidly. As the video images are stored on a flat disc and are recovered using controlled position laser heads, the still frame capability of a video disc is vastly superior to that of conventional VCR which has to continually scan the tape for still image reproduction. This has led to the development of integrated systems using computers and video disc players which can combine graphics, stills, full motion video and high quality audio to produce a very powerful tool for education and information systems.

The computer also enables the user to interact with video and graphic material in such as way as to provide attractive self-paced learning and information tools. A rapid search facility provides opportunities to construct dynamic video packages which suit the requirements of the individual user. Current developments are extending these capabilities, using video compression to make possible the storage of full motion video on computer hard disks or on compact discs. The field of video storage and retrieval is far from a closed book and those who are interested in the technology will continue to be excited for some time to come.

Mathematical quantities

ω	angular frequency
λ	record wavelength
$d\phi/dt$	rate of change of magnetic flux
Δ	frequency component error
Δf_m	maximum frequency deviation (TV)
ΔF_m	maximum frequency deviation (VCR)
θ	angle
τ	time period
δ	error
$\mathbf{B}$	magnetic field (vector)
$\mathbf{I}$	current (vector)
C	chrominance signal
V	video signal
Y	luminance signal
$\frac{1}{2} V_D$	frame reference signal
a	relative frequency deviation
$'A$	distance between full erase head and point where head starts to contact the tape
B_r	magnetic intensity
C_L	baseband colour signal
e	e.m.f. of the playback signal
E_c	Schmitt trigger power supply voltage
E_0	reference comparison voltage
f	signal frequency or rotational speed
F_0	carrier frequency
f_b	blank frequency
f_c	carrier oscillator frequency

f_H	horizontal frequency
f_m	maximum signal frequency
f_{max}	maximum recordable frequency
f_{0A}	channel A local oscillator frequency
f_{0B}	channel B local oscillator frequency
f_S	synchronizing tip frequency
f_{SC}	subcarrier frequency
$f_{S0(A)}$	channel A field signal frequency
$f_{0C(B)}$	channel B field signal frequency
$f_{0C(L)}$	downconverted C-signal subcarrier frequency
$f_{0C(L)A}$	downconverted channel A field signal frequency
$f_{0C(L)B}$	downconverted channel B field signal frequency
f_v	frame frequency (PAL)
f_w	white peak frequency
g	gap width of the video head
G	guard band width
h	height of RE above video head
H	period of the horizontal synchronizing pulse
i	recording current
I_0	collector current
I_c	Emitter current
I_m	recording current maximum amplitude
K	constant of proportionality
L	azimuth loss
m_f	modulation index
n	integer
N	number of coil turns on the video head
P	video track pitch
r	mounting radius of magnets
R	drum radius
R_H	Hall coefficient
S	synchronizing pulse
t	time
T	period of the recording signal
T	video track width
T_v	frame period (PAL)
V	head-to-tape speed
V_1	differential amplifier input in phase
V_2	differential amplifier input in anti-phase
V_3	base voltage
V_{be}	base-emitter voltage
V_{CC}	circuit supply voltage
V_{DD}	supply voltage (drain)
V_h	linear speed of the video head

V_H	Hall voltage
V_o	differential amplifier output voltage
V_{REF}	reference voltage
V_{SS}	supply voltage (source)
v_t	tape speed
Z	impedance
A	address
C	capacitance
D	diode
H	Hall element
IC	integrated circuit
LV	variable inductance
Q	transistor
R	resistor
RF	radio frequency
RV	variable resistor
SW	switch
TP	test point

Subscripts

ϕ	phase
Ω	angular frequency
A	head A
B	head B
c	carrier
CC	collector supply voltage
C1, ..., C4	collector 1, ..., collector 4
D	drum
d.c.	direct current
DD	drain supply voltage
e	emitter
e1, ..., e4	emitter 1, ..., emitter 4
E0	sequential number 0, 1, 2, 3, ...
F	sequential number 0, 1, 2, 3, ...
H	horizontal
L	low
(L)	low (modifier)
m	modulating
0 (zero)	zero
O	nominal

o	out
p	error code weighting (p)
q	error code weighting (q)
REF	reference
SC	subcarrier
SS	source supply voltage
v	video

Abbreviations

−(H)	high
−(L)	low
A/D C	analogue to digital converter
AB	address bus
ACC	autochromatic control
ACLK	clock terminal — memory controller
ACR	audio cassette recorder
ADJ	frequency deviation adjust
AFC	automatic frequency control
AFM	audio FM
AFT	auto frequency tracking
AGC	auto-gain control
AM	amplitude modulation
AND	logical AND
APC	auto phase control
APCM	audio pulse code modulation
AREC	audio record
ASS	assemble edit mode
AST	auto scanning track
ATBE	absolute time-base error
ATF	auto tracking finder
B/W	black and white
BET	surface area per gram of magnetic particles
BGH	sub format of PAL
BPF	bandpass filter
BS	control terminal on the RF converter
C	capacitor
CAS	column address strobe
CB	control bus

CH-A	channel A
CH-B	channel B
CLK	clock
CNR	chroma noise reducer
COMP	compensating
CPU	central processing unit
CRC	cyclic redundancy code
CTC	counter/timer controller
CTL	control signal
CV	capacitor in tuning circuit
D	diode
D/A C	digital to analogue converter
dB	decibel
DB	data bus
DG	differential gain
DM	demodulation
DOC	drop-out compensator
DP	differential phase
DRAM	dynamic random access memory
DTBC	digital time-base corrector
DTBE	differential time-base error
DTF	dynamic tracking finder
FF	fast forward
FG	frequency generator
FM	frequency modulation
FWD	forward
GC	gain control
GND	ground
H-SW	head switch
H-sync	horizontal synchronizing
HEM	Hall element memory
HES	Hall element switch
HF	high frequency
HPF	high pass filter
I/O	input/output
IC	integrated circuit
ID	identical detector
IF	intermediate frequency
INS	insert editing mode

INT	interrupt
IRQ	interrupt request
K REF	frame reference signal
LDR	light dependent resistor
LED	light emitting diode
LP	long play
LPF	low pass filter
LSI	large scale integration
LV	inductor in tuning circuit
MCU	monolithic processor
ME	metal (evaporated)
MIC	microphone
MM	monostable multivibrator
MOA	main operating amplifier
MP	metal (polished)
N-pole	north pole
NMOS	n type metal oxide semiconductor
NR	noise reduction
NTSC	North American Telecommunication Standards Committee
OE	output enable
OR	logical OR
OSC	oscillator
PB	playback
P/S	parallel/serial
PAL	phase alternate line
PCM	pulse code modulation
PG	pulse generator
PLL	phase-lock loop
PWM	pulse width modulation
Q	transistor
QV	tuning amplifier
R	resistor
R-S trigger	reset-set trigger
R/L	right/left
R/P	record/play

RAM	random access memory
RAS	row address strobe
RE	rotary erase
Re-MM	rotary erase monostable multivibrator
RE/WE	read/write
REC	record
REF	reference
REV	reverse
REW	rewind
RF	radio frequency
RF-SW	radio frequency switch
RF converter	radio frequency converter
RF signal	radio frequency signal
ROM	read only memory
RST	reset signal
RV	resistor (variable)
S	sync pulse
S-pole	south pole
S-side	supply side
S/P	serial/parallel
SAW	surface acoustic wave
SBI	serial bus input
SBO	serial bus output
SBT	serial bus trigger
SC	synchronizing
SECAM	Sequential Coleur À Memoiré
SIRQ	system interrupt request
SP	standard play
SPG	servo pulse generator
SPPG	speed pulse generator
SW	switch
SYNC	synchronizing signal
synch-separator	synchronizing separator
synch-tip	synchronizing tip
SYS CLR	system clear signal
SYSCON	system control
T	transistor
T-side	take-up side
TBC	time-base correction
TBE	time-base error
TENREG	tension regulator
TP	test point

TSB	trigger signal B
TSS	tilting sendust spatting
UHF	ultra high frequency
V	vertical
V-BLK	vertical blank
V-I	voltage to current
V-OSC	field oscillator
V-SYNC	vertical sync
VCA	voltage controlled amplifier
VCO	voltage controlled oscillator
VCR	video cassette recorder
VHF	very high frequency
VHS	video home system
VTR	video tape recorder
VXO	crystal-controlled oscillator
XOR	logical exclusive OR
Y channel	Y signal channel

Chapter 1

Introduction

1.1 What is a VCR?

A video cassette recorder (VCR) is a piece of equipment used to record and play back a video signal using magnetic tape as a storage medium. The VCR is a general storage device which has the capability to both record (write in information) and play back (read out information).

There are many storage devices such as a computer with memory, film projector, slide projector and audio cassette recorder (ACR), so it is necessary to add further conditions of definition to differentiate the VCR from these other forms of storage media.

1. The VCR is used to store principally video signals and not just audio, as in the ACR, or optical signals, as in the film projector.
2. The VCR uses magnetic cassette tape as a storage or recording medium and does not use semiconductor memories for the storage of signal information as in a computer, or discs as in the case of the laser disc system.

Therefore a VCR could be defined as a record/playback system which uses magnetic tape, in cassette form, as a storage medium with the sole purpose of recording or playing back video signal information. VCRs could be classified according to their quality and price in the world market. There are principally three kinds of VCR according to this classification.

1. *VCR of broadcast quality* These VCRs are used mainly by broadcast television companies and are designed and manufactured to the highest technical specifications. They are usually the most expensive.
2. *VCR of domestic quality* Since this type of VCR is manufactured to the minimum acceptable standards for the low cost volume market it is used mainly for domestic purposes. Examples of this type of recorder are the VHS (video home system) and Betamax format machines produced by the Japanese bulk market industries.

3. *VCR of professional quality* This type of recorder is used, for example, mainly in universities and media service centres. A typical example of this type of recorder is the Japanese U-matic which has a price and quality midway between the domestic and broadcast VCRs.

In addition to the VCR, users requiring high quality reproduction sometimes use a recorder which does not use the magnetic media in cassette form but as open reel tapes. These are known as video tape recorders or VTRs. The VCR (or VTR) may be placed in four further categories according to the width of magnetic tape or cassette used. These are:

1. A 2-inch-wide tape used for VTRs of broadcast quality.
2. A 1-inch-wide tape used for VTRs of broadcast quality.
3. A $\frac{3}{4}$-inch-wide tape cassette used for VCRs of broadcast and professional quality.
4. A $\frac{1}{2}$-inch-wide tape cassette used for VCRs of domestic quality.

Additionally, a cassette ($95 \times 625 \times 15$ mm) using a tape width of 8 mm has recently been used in the world market for a new type of VCR. This is similar to the size of cassette commonly found in audio cassette recorders (ACRs).

Video cassette recorders may also be classified according to the number of video heads used. These are:

1. Four heads for broadcast quality recorders using 2-inch tape.
2. Two heads used widely in VCRs using 1-inch, $\frac{3}{4}$-inch and $\frac{1}{2}$-inch tapes and cassettes.
3. One head, formerly used in VTRs but rarely manufactured now.

Lastly, classification may be determined by the way in which the heads are rotated in order to scan the magnetic tape pattern, as shown in Figure 1.1.

1. *Vertical scanning* The video heads move vertically to scan vertical magnetic tracks on the transporting tape. This method is used on four-head VCRs using 2-inch tape only.
2. *Helical scanning* The video heads move at an angle of less than 10°, relative to the transporting tape. Most VCRs use this format. Different types of VCR use various scanning angles.

1.2 Magnetic recording and playback principles

In terms of magnetic recording and playback principles, the video and audio cassette recorders are broadly similar. The significant difference is in the type of information recorded and how the information is recorded.

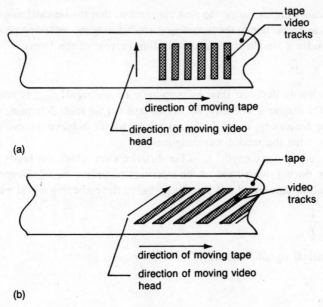

(a)

(b)

Figure 1.1   Two scanning methods in VCRs: (a) vertical scanning, (b) helical scanning.

The recording principle uses an electrical signal to vary a magnetic field which aligns magnetic material on a recording tape. In a magnetizing curve, as shown in Figure 1.2, the recording current (i) in the coil of a video head is transformed to magnetic intensity (B_r) on the tape. The relationship between recording current and magnetic intensity in the video head coil has some non-linear properties. The system is designed so that the signal operates in the linear region (i_1 to i_4) to ensure that the process of transformation from electrical to magnetic signal is linear.

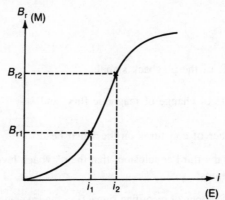

Figure 1.2   Magnetization curve.

We can therefore draw the first conclusion, that the residual magnetic intensity is in proportion to the recording current during the recording process.

Consider a simple sinusoidal recording current of the form:

$$i = I_m \sin \omega t \tag{1.1}$$

We have already deduced that the amplitude of the signal, I_m, is linearly related to the magnetic intensity, B_r, on the tape but we must determine what the angular frequency, ω, of the signal relates to. To achieve this we must first define what the record wavelength is.

The record wavelength, λ, is the distance over which the heads scan on the tape during one period of the electrical recording signal. Supposing v is the head-to-tape speed and T is the period of the recording signal where:

$$f = \frac{1}{T} \tag{1.2}$$

is the electrical signal frequency. Then:

$$\lambda = vT$$

or

$$\lambda = \frac{v}{f} \tag{1.3}$$

So the conclusion may be drawn that if the head-to-tape speed is kept constant the record wavelength will be inversely proportional to the electrical signal frequency.

The electrical recording signal is related to the magnetic signal through the video head scan on the tape. The playing back process obeys the laws of electromagnetic induction in which the e.m.f. of the playing back signal is in proportion to the rate of change of magnetic flux on the tape. So:

$$e = N \frac{d\phi}{dt} \tag{1.4}$$

where:

e = e.m.f. of the playback signal;

$\dfrac{d\phi}{dt}$ = rate of change of magnetic flux; and

N = number of coil turns on the video head.

So we can draw the third conclusion, that the playback level depends on the following conditions:

1. The rate of change of recording signal (i.e. the frequency of the recording

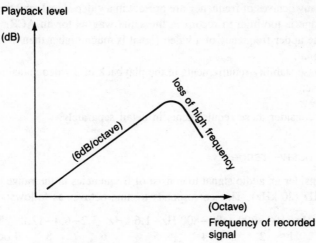

Playback level

(dB)

loss of high frequency

(6dB/octave)

(Octave)
Frequency of recorded
signal

Figure 1.3 Playback characteristic.

signal). For example, the playback level is zero if direct current was recorded.

2. The speed with which the head scans on the tape. Note: The scanning speed, or head-to-tape speed, is a combination of head and tape speed. Since the head is fixed in an ACR, the scanning speed is the tape speed. In a VCR the heads are rotating so it is possible to play back a still image from the tape in the pause mode. This is not possible in an ACR because of the fixed playback head. The playback characteristic of a video head is shown in Figure 1.3.

Normally the playback level is proportional to the frequency of the recorded signal, so a doubling of the recorded signal frequency results in a doubling of the playback level. Since a doubling of voltage level is 6 dB and a doubling of frequency is an octave, the playback characteristic curve is also known as a 6 dB/octave curve.

1.3 Practical requirements for recording a video signal

The principal difference between a VCR and an ACR has been defined as the fact that a VCR records mainly a video signal while the ACR records audio signals only. There are, however, significant differences in the way the signals are recorded in each case. These differences stem from the practical requirements for recording a video signal.

1. Many octaves of frequency are present in a video signal, and the octave range is too high to record in the same way as for an ACR.
2. The upper frequency of a video signal is much higher than in an audio signal.
3. Phase stability requirements in the playback of a video signal are very strict.

We can consider these requirements in detail separately.

A high octave range

If we consider an audio signal to consist of frequencies in the range of 50 to 20,000 Hz (20 kHz). This corresponds to nine octaves as follows:

50 Hz−100−200−400−800 Hz−1.6 kHz−3.2−6.4−12.8−25.6 kHz
 1   2   3   4   5   6   7   8   9 octaves

According to the playback characteristics of the head, a change of one octave results in a 6 dB change in playback level. So, an audio signal in the range 50 Hz to 20 kHz played back would result in a 54 dB change in playback level.

An input signal with all frequencies in the audio range 50 Hz to 20 kHz, of the same amplitude before recording, would result in a difference in playback level of 54 dB between the lowest and higher frequencies. The response, shown in Figure 1.4, is an undesirable, but inescapable, frequency distortion.

Usually, an equalizing network in the playback channel will be used to compensate for this frequency distortion. This response is shown in Figure 1.5. The frequency response of the network is the inverse of the playback characteristic, so that the output from the equalizing network is the same level as the signal recorded. In general, this type of compensation may be used to recover signals suffering from frequency distortion over a 70 dB range. However, the range of frequencies in a video signal is from 25 Hz (the frame frequency) to 6.5 MHz in the PAL (phase alternate line) system. An upper frequency limit of 3 MHz is usually used in a VCR of professional quality.

Using the same calculation as for the ACR, the octave range for a video signal of frequencies 25 Hz to 3 MHz, is 17 octaves. This would amount to a difference in playback level of 102 dB, far in excess of the 70 dB which can be compensated for using an equalizing network. To solve this problem, a recording scheme which involves the frequency modulation (FM) of the recording system has been adopted for all VCRs. Adopting FM moves the range of frequencies of the video signal and results in a reduction of the required number of octaves.

As is well known, the process of FM is the process in which the modulating signal is 'riding' on a carrier signal. Figure 1.6b shows the frequency spectrum of the modulated signal. The frequency of the carrier is in its original position and the frequency spectrum of the modulating signal

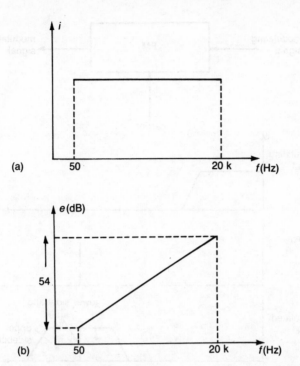

Figure 1.4 Effect of playback characteristic on the playback signal: (a) before recording, (b) after playing back.

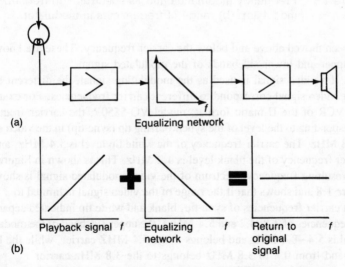

Figure 1.5 Equalizing network compensation for playback characteristic of the head in the playback channel: (a) PB channel in an ACR, (b) process of signal equalization.

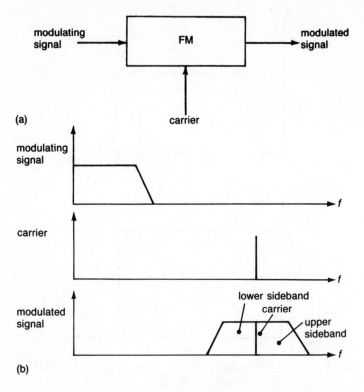

Figure 1.6 Frequency modulation and its spectrum: (a) frequency modulator, (b) range of frequencies in modulator.

has been moved above and below the carrier frequency. These are known as the upper and lower sidebands of the modulated signal.

If a video signal is used as the modulating signal, the different levels of the video signal correspond to different carrier frequencies. For example, in a VCR of the U-matic format, model VO-5850P, the carrier frequency corresponding to the level of the synchronizing tip (sync-tip) in the video signal is 3.8 MHz. The carrier frequency of the white tip level is 5.4 MHz, and the carrier frequency of the blank level is 4.2 MHz. This is shown in Figure 1.7. The resulting frequency spectrum of the video modulated signal is shown in Figure 1.8 and shows that if the range of the video signal is limited to 3 MHz (with carrier frequencies of sync-tip, blank and white tip indicated separately by frequencies of 3.8, 4.2 and 5.4 MHz), the upper sideband of the modulated signal is 5.4–8.4 MHz and belongs to the 5.4 MHZ carrier, while the lower sideband from 0.8 to 3.8 MHz belongs to the 3.8 MHz carrier.

As can be seen in Figure 1.8, the range of frequencies in the modulated video signal, which is called the RF (radio frequency) signal, is 0.8–8.4 MHz and corresponds to only four octaves. Therefore, it is important to note that

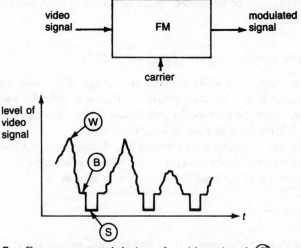

Figure 1.7   Frequency modulation of a video signal: (S) sync-tip
3.8 MHz; (B) blank 4.2 MHz; (W) white tip 5.4 MHz.

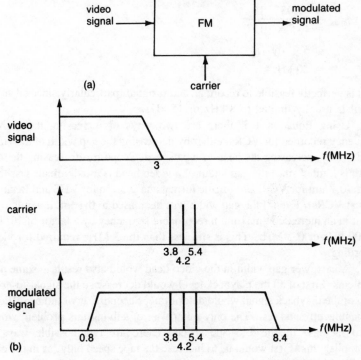

Figure 1.8   Frequency modulation of a video signal: (a) frequency
modulator, (b) frequency range in a frequency
modulated video signal.

a video signal must be modulated before recording. This is necessary for all kinds of VCR.

Video has a much higher upper frequency

It has been shown that while the frequency range of an audio signal to be recorded on an ACR is less than 20 kHz, the frequency range for a video signal to be recorded may be theoretically greater than 6 MHz, although this is usually less than 3 MHz for a professional quality VCR.

According to Equation 1.3 the highest recordable frequency is determined by the gap width of the head. This occurs when the wavelength is the same as the gap width of the head (g), i.e. when $\lambda = g$ or

$$f_{max} = \frac{v}{g} \tag{1.5}$$

For an ACR, the audio head is fixed and the head-to-tape speed (v) is equal to the tape speed (v_e). Generally in an ACR, $g = 5$ μm and $v_c = 9.05$ cm/s, and the highest recordable frequency (f_{max}) is therefore:

$$f_{max} = \frac{9.05 \times 10^{-2}}{5 \times 10^{-6}}$$

$$= 2 \times 10^4 \text{ Hz}$$
$$= 20 \text{ kHz}$$

So it is perfectly feasible to record an audio signal, particularly since an audio signal is usually limited to 8 kHz or 15 kHz.

Using Equation 1.5, there are two ways of increasing the highest frequency recorded for VCRs, either by decreasing the gap width of the video head, or by increasing the head-to-tape speed. Scope for decreasing the gap width is limited since the gap width of a video head is usually made less than 1 μm (0.7 μm for VCRs of U-matic format and 0.3 μm for VHS and Betamax format VCRs). Even if the gap width was decreased to 0.1 μm it would only result in an increase in maximum recordable frequency by a factor of 10, i.e. to 200 kHz or 0.2 MHz. This is still less than the 3 MHz required for video recording.

A narrower gap width in the video head would also result in some new problems. First of all the playback level would decrease so the signal-to-noise ratio of the playback signal would deteriorate. Secondly, it would be difficult to manufacture and test. The only other way of solving this problem would be to increase the head-to-tape speed. There are two possible ways to accomplish this. First would be to increase the tape speed only, as in the case of a fixed head ACR, with the second option to adopt a rotating video head to increase the head-to-tape speed v. We know that $f_{max} = 3$ MHz and $g = 1$ μm for a VCR, then:

$$v = v_t = f_{max} \times g$$
$$= 3 \times 10^6 \times 1 \times 10^{-6}$$
$$= 3 \text{ m/s}$$

resulting in a tape speed of 3 m/s, compared to the 9 cm/s in an ACR. So it is not practical to use a fixed video head and increase the tape speed. It would require an 11 km length of tape to play back for one hour. (You can imagine the weight and volume of a VCR using this format!) It is therefore important to consider using a rotating video head instead of a fixed head.

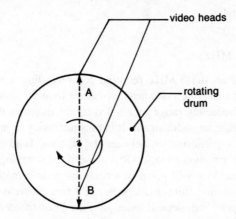

Figure 1.9   Rotating drum of a U-matic VCR: two video heads are fixed on the drum at 180° apart; the rotating speed of the drum is 25 turns/s; and the drum diameter is 110 mm.

Let us consider a VCR with U-matic format as an example, as shown in Figure 1.9. There is a rotatable head drum inside the VCR with two video heads fixed on the outside of the drum 180° apart. The rotational speed of the drum is 25 turns/s. In this way there are 50 tracks recorded on the tape per second with two tracks recorded per revolution of the drum. Because there are 50 fields, or 25 frames, per second in a video signal, one track corresponds to a field of the video signal and one frame of the picture will be recorded for each rotation of the drum. Given that the drum diameter, in this case, is 110 mm we can calculate the linear speed of the video head (v_h).

$$v_h = \omega R = \frac{2\pi f\phi}{2}$$
$$= \frac{2 \times 3.14 \times 25 \times 10^{-3}}{2}$$
$$= 8.64 \text{ m/s}$$

where: $\omega = 2\pi f$ is the angular frequency of the rotating drum;
$\phi = 2R$ is the drum diameter (R = radius of drum); and
f = rotational speed of drum in turns/s.

As stated earlier the head-to-tape speed is synthesized from head speed and tape speed. Because $v_t \ll v_h$ we can use v_h as an approximation for head-to-tape speed (v). Thus:

$$f_{max} \simeq \frac{v_h}{g}$$

$$= \frac{8.64}{1 \times 10^{-6}}$$

$$= 8.64 \text{ MHz}$$

which is greater than the 3 MHz required for recording a video signal. Therefore adopting a rotating video head provides a basis for video recording.

The higher frequency range of a video signal indicates that it carries more information than an audio signal. It is thus necessary to increase either the speed of accessing information between the tape and head or the density of storage of the information on tape. Obviously, if the scanning speed of the video head is increased it will result in writing more information on to, or reading more information from, the tape per unit time. Increasing the head-to-tape scanning speed will depend mainly on the head speed (v_h). It is not sufficient just to contain more information from the video signal if the horizontal scanning recording scheme is used, as in the ACR, because most of the recordable area on the tape is wasted. The recording density will be greatly increased if a helical scanning scheme is adopted.

The areas of tape which are recorded using a fixed head and helical scanning methods are shown in Figure 1.10 and indicate clearly that greater areas of the tape are used in the helical scanning scheme. A rough estimate of the amount of tape required to record one hour using the helical scanning method is readily possible. If the drum diameter is 110 mm and it takes the drum 20 ms to rotate half a turn (rotational drum speed is 25 turns/s) the length of track left by the scanning video head during the 20 ms is 0.17 metres. And so the track length scanned by video heads for one hour is approximately 30.6 km if placed end to end.

Phase stability

There is no problem with phase variations in the recorded signal for an audio recording but phase stability requirements for a video signal recording are very strict. The term 'phase of a video signal' means the periods of the horizontal synchronizing and the chromatic subcarrier signals. It is necessary to identify these periods accurately in both recording and playback. Because the played

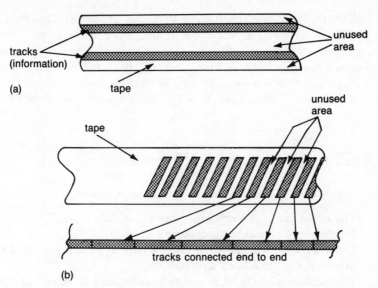

Figure 1.10 Scanning recording schemes of fixed or rotating heads: (a) horizontal scanning recording, (b) helical scanning recording.

back video signal originates from the recorded signal, through the head-to-tape scanning and electrical to magnetic signal transformation, the identity of the phase in REC (record) and PB (playback) obviously depends on the head-drum scanning system and the tape-transport system, i.e. the physical mechanism of the VCR. The relative difference in time between the PB signal and the REC signal is known as the time-base error (TBE). This is a very important concept in video recording. It is very difficult to eliminate TBE completely but there are a number of measures which can be taken to minimize it. The accuracy of the drum assembly and capstan shaft system can be manufactured to an accuracy of 1 μm or less. Additionally, the drum motor and capstan motor can be controlled through separate drum and capstan servos. In general, adopting these measures will provide a satisfactory solution for black and white (B/W) images, but it is not enough for colour pictures because the period of the subcarrier is 283 times less than that of the horizontal synchronizing (H-sync) signal. Thus, a steady colour image cannot be achieved unless other methods are adopted.

A colour video signal (V) consists of the luminance signal (Y) and a chrominance signal (C) such that:

$$V = Y + C \tag{1.6}$$

These can be separated from the video signal and the C signal can be downconverted. The frequency subcarrier (f_{SC}) results in downconversion from 4.43 MHz to under 1 MHz for the C signal. For example, this is as low as 0.685 MHz in the case of a U-matic VCR type VO-5850P. The effect of TBE will be improved as the f_{SC} is shifted down in frequency. This, unfortunately, is still not sufficient for colour signal recording. Another important scheme for improving the effects of TBE is the adoption of time-base correction (TBC) circuits. These will be discussed in detail later. In general, adoption of these methods will result in a steady colour picture.

1.4 The features of a VCR

The features of a VCR can be briefly described as follows:

1. The signal has to be modulated before recording.
2. Rotating video heads are necessary.
3. Drum servo and capstan servo systems have to be used to control the drum and capstan motors separately.
4. The mechanism of the VCR is to be manufactured to better than 1 μm accuracy.
5. For VCRs using $\frac{3}{4}$- and $\frac{1}{2}$-inch tape, the chromatic signal is separated from the video signal and downconverted using a subcarrier before recording in the REC C channel. The PB C signal (i.e. $f_{SC(L)}$) is converted up in frequency and is combined with TBC and returned to the normal subcarrier (f_{SC}).

We can consider these features using a VCR of the U-matic format as a basis for discussion.

The colour video signal (V) is separated into the luminance signal (Y) and the chrominance signal (C). The Y signal is modulated up in frequency and the C signal is downconverted in frequency. A U-matic format VCR of type VO-5850P downconverts the C signal from 4.43 to 0.685 MHz and a VHS downconverts the C signal to 0.627 MHz. Therefore, an REC channel consists of both a Y and a C channel in which the Y signal passes through a frequency modulator and is converted to an RF signal (i.e. modulated signal), while the C signal passes through a frequency converter and is converted down to subcarrier ($f_{SC(L)}$) in the REC C channel, where:

$$f_{SC(L)} = f_C - f_{SC} \qquad (1.7)$$

f_C is the frequency of an oscillator in the VCR. The RF signal and $f_{SC(L)}$ are combined and sent to the head for recording.

In the PB channel the playback signal from the head is separated into the RF signal and $f_{SC(L)}$ signal first, and the RF signal is demodulated and converted back to the Y signal in the PB Y channel. The subcarrier frequency $f_{SC(L)}$ passes through a frequency converter to return it to the subcarrier frequency f_{SC}. Time-base correction is carried out at this point. The

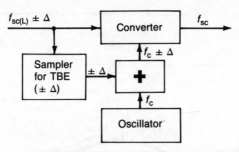

Figure 1.11 The method of TBC in PB C channel.

recombined Y and C signals are output from the VCR. The method of TBC in the PB C channel, mentioned earlier, is shown in Figure 1.11.

The PB C signal with TBE ($\pm \Delta$) is indicated by $f_{SC(L)} \pm \Delta$. The TBE is sampled and then mixed with a local oscillator frequency f_C to produce $f_C \pm \Delta$. Both $f_{SC(L)} \pm \Delta$ and $f_C \pm \Delta$, which have the same TBE, are frequency converted and the error, $\pm \Delta$, is cancelled during the conversion process. That is:

$$(f_C \pm \Delta) - (f_{SC(L)} \pm \Delta) = f_{SC} \tag{1.8}$$

This correction of TBE, known as pseudo-TBC, is in the C channel only and does not include the Y channel. The key systems and main parts of a VCR are shown in Figure 1.12. These are comprised of two channels, audio and video. Each channel includes PB and REC channels. The audio head is fixed and the video heads (usually two) are rotatable. The drum motor and capstan motor are controlled separately by the drum servo and capstan servo systems which drive the drum and capstan shaft respectively.

The pinch roller fits tightly against the capstan shaft to transport the tape, while the drum drive rotates the video heads to scan the transport tape. Synchronized pulses separated from the input video signal are fed to the framing circuit and are transferred to the reference signal after the framing process. This reference signal is compared with the sampling signals from the rotating speed and phase of the drum and capstan motors in their respective servo systems, and further synchronize scanning the track with the synchronizing pulse. This relates the magnetic signal to the recording electrical signal and is called the recording phase. The reel motor drive supply or take-up reel, through the reel idler, ensures take up of the transporting tape. A reel servo controls the reel motor to ensure correct tension of the tape as it is transported.

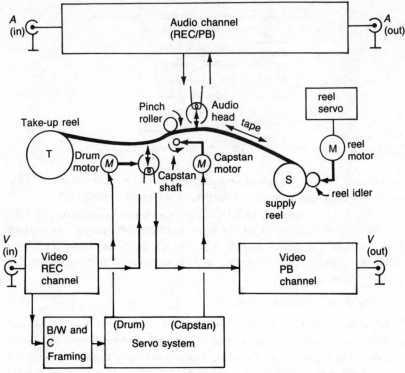

Figure 1.12 Configuration of the key systems and main components in a VCR.

Chapter 2

Basic principles used in the VCR

2.1 Frequency modulation (FM) and demodulation (DM)

2.1.1 The FM characteristic

The frequency modulator characteristic is shown in Figure 2.1, the slope of which ensures that the input level of the video signal corresponds to the required frequency. The changing level of the input signal to the modulator results in a corresponding frequency deviation. The sync-tip, blank and white peak level in the video signal correspond to f_s, f_b and f_w after modulation.

2.1.2 Bandwidth of the recorded signal

Because the bandwidth of the recorded signal is limited by the playback characteristic of the video head, there are several differences in the way the signal has to be modulated and demodulated in a VCR. For illustration this can be compared with the frequency modulated sound signal in television.

Carrier frequency too low for modulation

In television	carrier frequency	6.5 MHz
	bandwidth of sound frequencies	20 kHz
In VCR	carrier frequency (e.g. VO-5850P)	3.8–5.4 MHz
	bandwidth of video frequencies	3 MHz

It is general practice in FM that the carrier frequency is ten times higher than the bandwidth of the modulating signal. This is to prevent a leakage of the

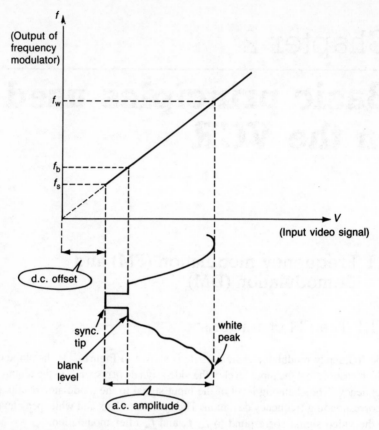

Figure 2.1 Frequency modulator characteristic.

modulating signal appearing at the output of the modulator and a leakage of the carrier appearing after demodulation. This is no problem in television but in a VCR the carrier is near the region of the video frequencies and it is necessary to include a suitable filter after the modulator or demodulator to prevent leakage occurring.

Modulating index is too small for modulation

In a television the maximum frequency deviation, ΔF_m, which is decided by the television station, is 50 kHz and the highest sound frequency, f_m, is 20 kHz. So if the modulation index, m_f, is given by

$$m_f = \frac{\Delta \omega_m}{\Omega} = \frac{\Delta f_m}{f_m}$$

where Δf_m is the maximum absolute frequency deviation the minimum modulating index for the sound signal in television is that $m_f > 2.5$. In a VCR the region of frequency deviation is $3.8-5.4$ MHz so

$$\Delta f_{\mathrm{m}} = \tfrac{1}{2}(5.4 - 3.8) = 0.8 \text{ MHz}$$

and the highest video frequency recorded is 3 MHz then

$$m_f = \frac{0.8}{3} = 0.27 \ (<1)$$

So it can be seen that the television sound signal modulation occupies six times more bandwidth than in VCR video modulation.

For $f_{\mathrm{m}}(n = 4)$ this is 120 kHz (or 0.12 MHz) and is still much less than the utilizable frequency region in television, 250 kHz or 0.25 MHz. However, the frequency region of the modulated video signal in VCR still exceeds the upper limit of the frequency range of the playback characteristic, even though it uses only a single pair of side frequencies ($n = 2$), so it is necessary to attenuate the amplitude in higher frequencies of the upper sideband in the modulated signal before recording.

The small value of m_f indicates that the signal is weak after demodulation and it is necessary to arrange suitable circuits for improving the signal-to-noise (S/N) ratio of the demodulated signal in the PB channel, e.g. the noise cancellation circuit, etc.

Relative frequency deviation for modulation is too high

The relative frequency deviation a is defined as the ratio of maximum absolute frequency deviation (ΔF_{m}) to carrier frequency (F_0), where

$$a = \frac{\Delta F_{\mathrm{m}}}{F_0} \qquad\qquad (2.1)$$

In television $F_0 = 6.5$ MHz and $\Delta F_{\mathrm{m}} = 50$ kHz $= 0.05$ MHz so

$$a = \frac{0.05}{6.5} \simeq 0.01$$

In VCR

$$F_0 = \tfrac{1}{2}(3.8 + 5.4) = 4.6 \text{ MHz}$$
$$\Delta F_{\mathrm{m}} = 0.8 \text{ MHz}$$

so

$$a = \frac{0.8}{4.6} = 0.16$$

which is much greater than 0.01 in television. In a tuned circuit

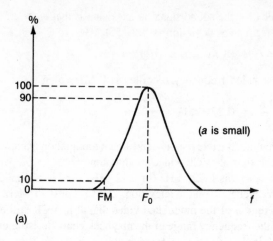

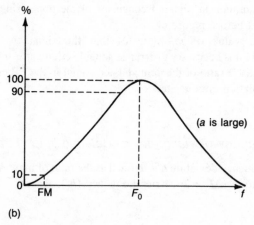

Figure 2.2 Detuning ratio.

$$a = \frac{\Delta F_m}{F_0} = \frac{F_m - F_0}{F_0}$$

indicates the detuning ratio as shown in Figure 2.2. Since a is large, the linear region in which the slope detector can operate is too narrow, therefore it is difficult to adopt slope modulation and demodulation as in television sound signal modulation.

2.1.3 Frequency modulator circuit used in VCR

A voltage crystal-controlled oscillator (VXO) is generally used as a kind of slope modulator to generate frequency modulated video. We have seen,

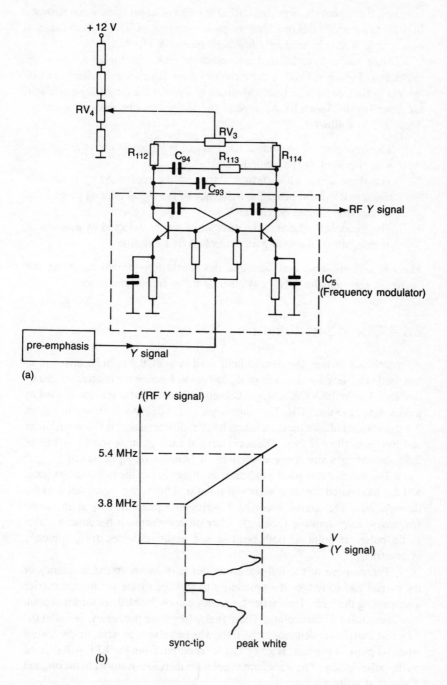

Figure 2.3 Frequency modulator in the VO-5850P.

however, that since a is large, the utilizable range of linear slope is too narrow. In order to get round this problem, the push—pull type of VXO, which consists of a double VXO, is used in a few high quality VTRs.

Most VCRs have utilized a monostable multivibrator as a frequency modulator. Figure 2.3 shows the circuit of the frequency modulator used in the VO-5850P. Basic principles and circuit analysis of the frequency modulator are described in Appendix A. Several conclusions can be drawn, however. These are as follows:

1. The frequency of the output RF Y signal is varied in direct proportion with the level of the input Y signal.
2. The slope of the characteristic curve can be controlled by a variable resistor (RV_4) and hence the frequency of the signal relating to the level of the sync-tip can be chosen.
3. The frequency related to the peak white can be selected by adjustment of the gain of the auto-gain control (AGC) amplifier.

The FM circuits described here and the conclusions drawn regarding the operation are valid for VHS and other U-matic format machines.

2.1.4 The demodulator

As mentioned earlier, the demodulator used in a VCR has to be different to that used in a television, i.e. a slope demodulator, because the relative frequency deviation (a) in the VCR is large. Usually a counter-pulse scheme is used as a demodulator in the VCR. The counter-pulse demodulator is shown in Figure 2.4 and consists of four parts including limiter, differentiator, full-wave rectifier and low pass filter (LPF). The waveform at each point is shown in Figure 2.4b, assuming a sine wave is used as an example of input signal.

The waveform at point a indicates the Y signal, i.e. the modulating signal, and the modulated signal is shown at point b. This is the signal fed into the demodulator. The signal becomes a series of square waves, at the same frequency, after limiting (point c). After differentiation it becomes a series of tip pulses at point d. Both positive and negative pulses are completely symmetrical.

The purpose of the full-wave rectifier is to move up the frequency of the carrier and to reduce the possibility of leakage of the modulated carrier after passing the LPF. The carrier frequency can be doubled, as shown at point e, without loss of information. The signal, after these processes, is fed to the LPF and charges capacitance in the filter when a pulse is present, or discharges when no pulse is present. High or low level output from the LPF will depend on the pulse density. The waveform at point f is therefore restored to the original Y signal at point a.

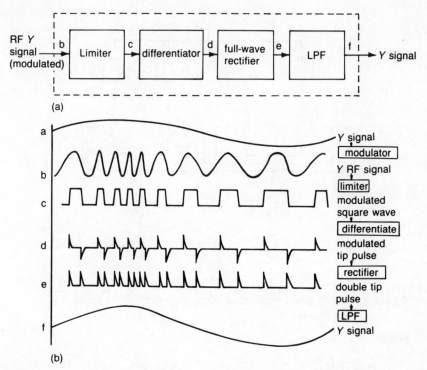

Figure 2.4   The counter-pulse demodulator and its waveforms.

2.2 The VCR chroma channel frequency converter

The frequency converter in a VCR uses a balanced modulator using transistors as a dual differential amplifier. If two differential amplifiers are connected as shown in Figure 2.5 we would obtain a balanced modulator. The features of this type of connection are that:

(a) V_1 is input to the two differential amplifiers in phase;
(b) V_2 is input to the two differential amplifiers in anti-phase; and
(c)   the output voltage V_O is contributed to by both collector currents from T_1 and T_4 ($I_{c1} \sim I_{e1}$ and $I_{c4} \sim I_{e4}$).

Using these features we can obtain the following results:

$$I_{e1} = I_O + I_\Omega - I_\omega + I_{\omega+\Omega} + I_{\omega-\Omega} \tag{2.2}$$

$$I_{e2} = I_O + I_\Omega + I_\omega - I_{\omega+\Omega} - I_{\omega-\Omega} \tag{2.3}$$

$$I_{e3} = -I_O - I_\Omega - I_\omega - I_{\omega+\Omega} - I_{\omega-\Omega} \tag{2.4}$$

$$I_{e4} = -I_O - I_\Omega + I_\omega + I_{\omega+\Omega} + I_{\omega-\Omega} \tag{2.5}$$

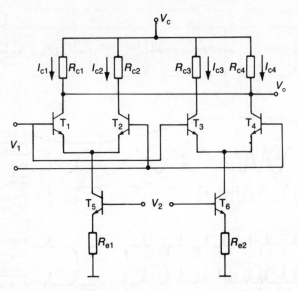

Figure 2.5 Balanced modulator using two dual differential amplifiers.

where

$$I_O = \tfrac{1}{2} \frac{B_0}{R_e}$$

$$I_\Omega = \tfrac{1}{2} \frac{B}{R_e} \cos \Omega t$$

$$I_\omega = 10 \frac{AB_0}{R_e} \cos \omega t$$

$$I_{\omega+\Omega} = 5 \frac{AB}{R_e} \cos (\omega+\Omega)t$$

$$I_{\omega-\Omega} = 5 \frac{AB}{R_e} \cos (\omega-\Omega)t$$

The derivation of these equations is shown in detail in Appendix B. Also

$$V_o = R_c (I_{c1}+I_{c4})$$
$$\simeq R_c (I_{e1}+I_{e4}) \tag{2.6}$$
$$= 2R_c (I_{\omega+\Omega}+I_{\omega-\Omega})$$

These expressions show that in the output signal of the balanced modulator, only the upper side frequency $(\omega+\Omega)$ and the lower side frequency $(\omega-\Omega)$ are produced. The main reason that the modulator is connected in this

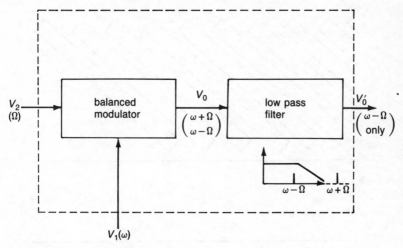

Figure 2.6 Frequency converter.

way is that the carrier component (ω) does not appear in the output. Using an LPF with the upper limit frequency set between ($\omega+\Omega$) and ($\omega-\Omega$) we can obtain the component ($\omega-\Omega$) from the input frequencies ω (i.e. V_1) and Ω (i.e. V_2). It is this type of frequency converter that is used in the VCR shown in Figure 2.6.

2.3 VHS high density recording techniques

2.3.1 Outline

With a conventional VCR system, such as U-matic format, there should be a guard band on the tape between the recorded video tracks, provided for the purpose of preventing crosstalk between adjoining tracks caused by deviation of the tracks due to slight variations between machines. If it is possible to realize a mechanism which will not generate crosstalk, even if track error happens to be caused by the tape pattern, the portion of the tape where guard bands are provided would become entirely unnecessary and it would become possible to reduce the length of the tape by eliminating the portion of the tape used for guard bands. In addition, it would also be possible to further shorten the length of the tape by reducing the width of the track itself, if the problems of S/N ratio in the PB mode could be solved.

Figure 2.7 shows the magnetic tape pattern tracks recorded by a conventional VCR (Figure 2.7a) and by a VHS recorder (Figure 2.7b). Data outlining the width of the video track (T), the guard band (G) and video track

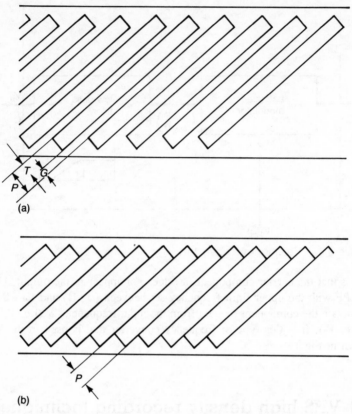

Figure 2.7 (a) Conventional video track pattern. (b) VHS video track pattern.

Table 2.1 Video track data comparisons

	NTSC			PAL		
	U-matic (H)	VHS	Betamax	U-matic	VHS	Betamax
Track width $T(\mu m)$	85	58	58	85	48.6	32
Guard band width $G(\mu m)$	52	0	0	80	0	0
Track pitch $P(\mu m)$	137	58	58	165	48.6	32

pitch (P) are shown in Table 2.1. It can be seen from the table that the video track pitch will be reduced and consequently the amount of tape will be reduced by about a third if:

(a)   the track width of the video head itself is reduced to reduce the width of the video track (T); and

(b)   the guard bands (G) are eliminated and the tape speed is decreased to reduce the track pitch (P).

Even if the amount of tape is reduced, miniaturization of the video deck will be limited unless the revolving drum is also miniaturized. As was shown in Chapter 1, reduction of the drum diameter results in a reduction of the relative speed of the head and tape (v) according to the following formula:

$$\lambda = \frac{v}{f}$$

The reduction in head-to-tape scanning speed (v) and the recorded wavelength (λ) on the tape will mean that the highest recordable frequency (f) on the tape will decrease and will result in a deterioration of the reproduced images.

From the characteristics of magnetic playback, the relationship between the minimum recorded wavelength on the tape and the gap width (g) of the video head can be determined from the following formula:

Table 2.2 Comparison data for PAL video recorders

	U-matic	VHS	Betamax
Tape speed (cm/s)	9.53	2.339	1.87
Gap width of video head (μm)	1.0	0.3	0.3
Drum diameter (cm)	110	62	74.5
Relative head-to-tape speed (m/s)	8.34	4.85	5.83
Recorded wavelength on tape (μm) (signal frequency 5 MHz)	1.8	0.97	0.97
Tape consumption (m/h)	6.5	1.52	0.9
Size of cassette (T) × (D) × (W) (mm³)	32 × 140 × 222	25 × 104 × 188	25 × 96 × 156

$$\lambda \leq \frac{g}{2}$$

Thus, it is necessary to make the gap of the video head narrower in order to decrease the relative speed of the head and tape and thereby to expand the range of frequencies to be handled into a high enough range.

Table 2.2 shows a comparison of specifications from a range of video recorders (for the PAL system only). As shown in the table, the tape speed in the VHS system has been reduced to about $\frac{1}{8}$ of that of the conventional U-matic type. The drum diameter has also been reduced to about $\frac{1}{2}$ of that in the U-matic system. With these improvements it has become possible to miniaturize the body of the deck and to record on the tape for a longer time.

As mentioned earlier, various problems can be expected from the adoption of this super high-density system, for example:

(a)   problems of interference between adjoining tracks arising from the elimination of the guard bands in the magnetic pattern; and

(b)   problems with the S/N ratio due to the reduced width of the video track and the relative speed.

2.3.2 Azimuth recording system

In conventional VTRs the head gap is positioned square to the direction of the video track. It is easily understood that the playback output may be considerably decreased if this angle varies between recording and playback. This is called azimuth loss (L) which can be expressed as:

$$L = -20 \log \left(\frac{\sin \beta}{\beta} \right) \text{ (dB)}$$

where

$$\beta = \frac{\pi T \tan \theta}{\lambda}$$

and T = track width;
  λ = recorded wavelength ($\lambda = v/f$); and
  θ = azimuth angle.

To minimize the problem of interference between adjacent tracks, an azimuth recording system has been used in the VCR, which also eliminates guard bands in the tape pattern. This is used in recording systems such as Betamax, VHS, 8 mm VCRs and VHS-C, etc.

In the VHS system, for example, the luminance signal is frequency modulated so that the end (sync-tip) of its synchronizing signal is 3.8 MHz and the white peak is 4.8 MHz. The converted chrominance carrier signal has been converted to 627 kHz. To overcome the interference between adjacent tracks, a total azimuth angle of 12° is provided by inclining the gaps of the CH-A and CH-B heads by 6°, respectively, in relation to the perpendicular of the video track, but in opposite directions to each other, as shown in Figure 2.8.

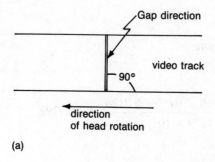

(a)

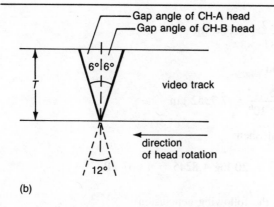

(b)

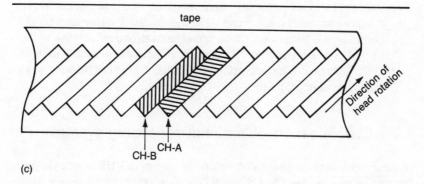

(c)

Figure 2.8   (a) Normal angle between track and gap. (b) Azimuth angle of VHS. (c) Azimuth recording in VHS.

In the case of VHS, the azimuth loss can be calculated as follows:

v = 4.85 m/s
T = 48.6 μm
θ = 12° and tan 12° = 0.2126
f = 4.3 MHz (average of 3.8 MHz and 4.8 MHz)

and

$$\lambda = \frac{4.85}{4.3 \times 10^{-6}} = 1.1279 \ \mu m$$

so

$$\beta = \frac{\pi T \tan \theta}{\lambda} = \frac{3.14 \times 48.6 \times 10^{-6} \times 0.2126}{1.1279 \times 10^{-6}} = 28.7792$$

$$\sin \beta = 0.4837$$

and

$$L = 20 \log \frac{28.7792}{0.4837} = 35.5 \ dB$$

Because f = 627 kHz and

$$\lambda = \frac{v}{f} = \frac{4.85}{0.627 \times 10^6} = 7.7352 \ \mu m$$

in the chrominance signal, then:

$$L = 20 \log \frac{\beta}{\sin \beta} = 20 \log 4.8245 = 13.67 \ dB$$

To sum up we can draw the following conclusion.

The problem of crosstalk generated by the luminance signal components can be minimized, but the crosstalk generated by the chrominance component in the signal has a strong influence and this cannot be readily solved by just using an azimuth recording system. The colour phase shift recording system has been introduced to solve this particular problem.

2.3.3 Chrominance phase shift recording system

Using the chrominance phase shift recording system in VHS as an example, the scheme may be described in the following way:

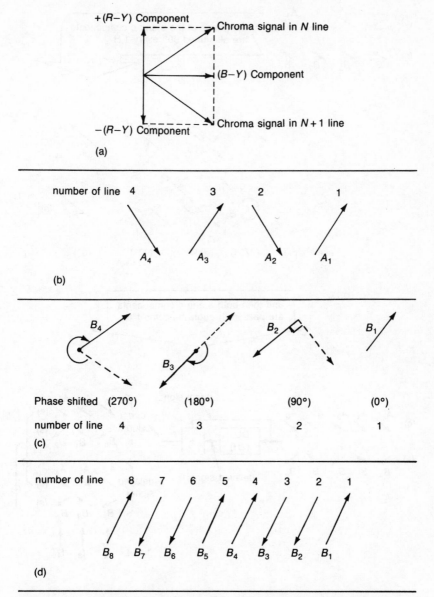

+(R−Y) Component

Chroma signal in N line

(B−Y) Component

−(R−Y) Component

Chroma signal in N + 1 line

(a)

number of line   4   3   2   1

A_4 A_3 A_2 A_1

(b)

B_4 B_3 B_2 B_1

Phase shifted   (270°)   (180°)   (90°)   (0°)

number of line   4   3   2   1

(c)

number of line   8   7   6   5   4   3   2   1

B_8 B_7 B_6 B_5 B_4 B_3 B_2 B_1

(d)

Figure 2.9   (a) The phase of the C signal in the PAL system alternates line by line. (b) The phase of the chroma signal recorded on the CH-A track. (c) The phase is shifted 90° every 1H and alternates line by line. (d) The phase of the chroma signal recorded on the CH-B track.

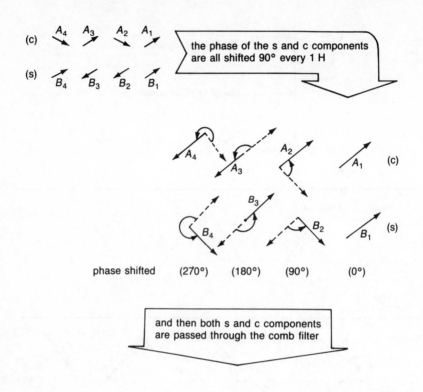

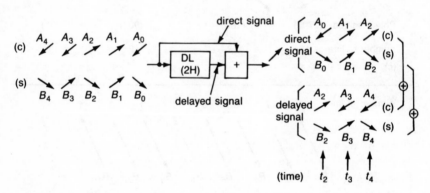

Figure 2.10 (a) Processing of the PB signal before filtering. (b) Processing the PB signal picked up from the CH-B head. The output from the adder is:

s component	$B_0 + B_2$,	$B_1 + B_3$,	$B_2 + B_4$	are doubled
c component	$A_0 + A_2$,	$A_1 + A_3$,	$A_2 + A_4$	are removed
	$\uparrow$	$\uparrow$	$\uparrow$	
time	t_2	t_3	t_4	

During recording: CH-A track records a chrominance signal with the phase un-shifted

CH-B track records a chrominance signal with the phase shifted by 90° every 1H (H = one field period).

During playback: CH-A playback signal from head A passes through a comb filter, which consists of a 2H delay line and an adder. The output signal component, from the CH-A track, out of the adder is doubled and the crosstalk component, i.e. the signal picked up from the CH-B track, is removed.

CH-B playback signal from head B is shifted by 90° every 1H before it is passed through the comb filter. The signal component from CH-B is doubled and the crosstalk component picked up from CH-A track is removed.

Figure 2.9 shows the phase of the chrominance signal recorded on the tape in the VHS system (PAL system only) and Figure 2.10 shows how the crosstalk component is removed using chrominance phase shift recording.

2.3.4 High density recording in other systems

It must be pointed out that the high density recording technique described in the example above is used in the VHS and PAL systems only. A VCR using a $\frac{1}{2}$-inch tape is slightly different.

1. To remove crosstalk generated by the luminance components in Betamax, a total azimuth angle of 14° is provided by inclining each of the CH-A and CH-B heads at an angle of 7°, not 6°, in relation to the perpendicular of the video track, and in the opposite direction to each other.

2. To remove crosstalk generated by the chrominance components in the VHS (NTSC) system, the phase of the signal for the CH-A field is shifted clockwise by 90° every 1H and the phase of the signal for the CH-B field is shifted counter-clockwise by 90° every 1H during recording. During playback the signal picked up by the CH-A head is shifted counter-clockwise by 90° every 1H before it passes through a comb filter consisting of a 1H delay line and an adder. Similarly, the signal picked up by the CH-B head is shifted clockwise by 90° every 1H before it passes through the same comb filter. The result is that the signal component output from the adder is increased by a factor of two and the crosstalk component is removed.

3. To remove crosstalk generated by the chrominance components in the Betamax (NTSC) system, CH-A records a chrominance signal in which the phase is un-shifted, while CH-B track records a chrominance signal with the phase shifted by 180°, i.e. inverted, every 1H during recording. The PB

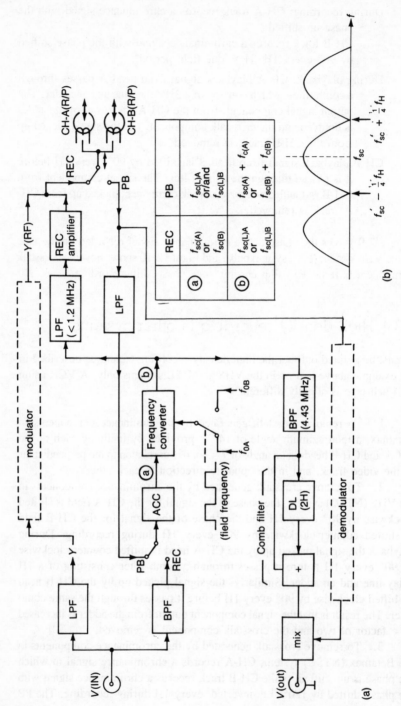

Figure 2.11 (a) Betamax (PAL system). (b) Characteristic of the comb filter.

signal from CH-A head is passed directly through a comb filter consisting of a 1H delay line and an adder, but the PB signal from CH-B head needs to be shifted 180° every 1H before it passes through the comb filter. The result is the same as in the case discussed above, i.e. the signal component is doubled and the crosstalk is removed.

4. To remove crosstalk generated by the chrominance components in the Betamax (PAL) system a scheme is adopted in which the subcarrier frequency is downconverted alternately between $(44+\frac{1}{8})f_H$ and $(44-\frac{1}{8})f_H$. It is necessary to arrange two local oscillators to carry this out, where:

$$f_{0A} = (284-\tfrac{1}{4})f_H + (44-\tfrac{1}{8})f_H$$

and $$f_{0B} = (284-\tfrac{1}{4})f_H + (44+\tfrac{1}{8})f_H$$

During recording the CH-A signal, $f_{sc(A)} = (284-\frac{1}{4})f_H$, will be mixed with f_{0A} and converted from $f_{sc(A)}$ to $f_{sc(L)A} = (44-\frac{1}{8})f_H$; the CH-B field signal, $f_{sc(B)} = (284-\frac{1}{4})f_H$ will be mixed with f_{0B} and converted from $f_{sc(B)}$ to $f_{sc(L)B} = (44+\frac{1}{8})f_H$.

During playback the PB signal picked up by the CH-A head is still mixed with f_{0A} and converted, i.e. the signal component is

$$
\begin{aligned}
f_{sc(A)} &= f_{0A} - f_{sc(L)A} = (284-\tfrac{1}{4})f_H + (44-\tfrac{1}{8})f_H - (44-\tfrac{1}{8})f_H \\
&= (284-\tfrac{1}{4})f_H
\end{aligned}
$$

while the crosstalk component is

$$
\begin{aligned}
f_{c(A)} &= f_{0A} - f_{sc(L)B} = (284-\tfrac{1}{4})f_H + (44-\tfrac{1}{8})f_H - (44+\tfrac{1}{8})f_H \\
&= (284-\tfrac{1}{4})f_H - \tfrac{1}{4}f_H
\end{aligned}
$$

In other words, there is a frequency interval of $\frac{1}{4}f_H$ between the signal and the crosstalk components in the PB signal picked up by the CH-A head. The PB signal picked up from the CH-B head is also mixed with f_{0B} and converted, i.e. the signal component is:

$$
\begin{aligned}
f_{sc(B)} &= f_{0B} - f_{sc(L)B} = (284-\tfrac{1}{4})f_H + (44+\tfrac{1}{8})f_H - (44+\tfrac{1}{8})f_H \\
&= (284-\tfrac{1}{4})f_H
\end{aligned}
$$

and the crosstalk component is

$$
\begin{aligned}
f_{c(B)} &= f_{0B} - f_{sc(L)A} = (284-\tfrac{1}{4})f_H + (44+\tfrac{1}{8})f_H - (44-\tfrac{1}{8})f_H \\
&= (284-\tfrac{1}{4})f_H + \tfrac{1}{4}f_H
\end{aligned}
$$

There is also a frequency interval of $\frac{1}{4}f_H$ between the signal and the crosstalk components in the PB signal picked up by CH-B head.

A comb filter, consisting of a delay line of 2H and an adder in which the frequency interval between the crest and trough in the characteristic is $\frac{1}{4}f_H$, can be used to suppress the crosstalk component. The scheme used in removing the crosstalk component in a Betamax (PAL) system is shown simply in Figure 2.11. The autochromatic control and frequency converter circuits are also shown in the diagram. During the recording the input signal is $f_{sc(A)}$ in the CH-A field and $f_{sc(B)}$ in the CH-B field. They are converted to $f_{sc(L)A}$ and $f_{sc(L)B}$ respectively, after they are mixed with f_{0A} or f_{0B}.

During playback $f_{sc(L)B}$ (or $f_{sc(L)A}$) appearing in the input signal is picked up from CH-A (or CH-B) head if crosstalk appears and they are converted, respectively, to $f_{sc(A)}+f_{c(A)}$, or $f_{sc(B)}+f_{c(B)}$, where there is a difference of $\frac{1}{4}f_H$ between $f_{sc(A)}$ and $f_{c(A)}$ or $f_{sc(B)}$ and $f_{c(B)}$. A comb filter which has the characteristic shown in Figure 2.11b is arranged to select the signal component, $f_{sc(A)}$ or $f_{sc(B)}$, and remove the crosstalk component, $f_{c(A)}$ or $f_{c(B)}$.

2.4 VCR servo system basic principles

Using a mechanism or a circuit to produce a signal which samples and feeds back a parameter of an object which is moving or varying, and comparing this signal with a reference signal, the object can be controlled by adjustments to the object in proportion to the difference in these signals. The moved or varied error can be limited to a specified range. This automatic control and feedback system is called a servo system.

There are three main kinds of servo systems used in a professional quality VCR:

- drum servo system
- capstan servo system and
- tension (or reel) servo system

In the drum or capstan servo system the sampling parameters are the rotating speed and phase of the drum or capstan motor. The reference parameter is the frame reference signal ($\frac{1}{2} V_D$). Using a phase-lock loop, the rotational speed and phase of the motor will be synchronized with the frequency and phase of the reference signal.

In the reel servo system, the sampling parameter is the tension on the transporting tape and the reference parameter is a pre-defined position of the regulator tension arm. A brake on the supply reel side or the take-up reel side ensures that an approximately constant tension of the tape is maintained. Because there is a large time constant in the motor, a long time is needed to restore the phase flutter if an interference appears in the phase of the reference signal. An auxiliary damper loop operates in conjunction with the main phase-lock loop to prevent interference effects of this kind. The phase-lock is known as the phase loop whilst the damper loop is known as the speed loop. The speed loop is not only used to suppress the interference but also to improve the starting performance and smooth running of the motor.

2.4.1 Servo system concepts

It is necessary to introduce some terms and concepts associated with the servo system before discussing the basic principles of the servo system itself.

The framing reference signal ($\frac{1}{2} V_D$) and the control signal (CTL)

A 25 Hz frame square wave is extracted from the input video signal in the REC mode, i.e. from the input field synchronizing signal, by the synchronizing separator. This is known as the frame reference signal ($\frac{1}{2} V_D$). However, in the PB mode there is no external synchronizing signal being input so a signal is divided from the internal crystal oscillator, frequency 4.43 MHz, to form the reference signal $\frac{1}{2} V_D$. The $\frac{1}{2} V_D$ is used as a reference signal by the drum and capstan servos to lock the rotating speed and phase of the drum and capstan motors to the $\frac{1}{2} V_D$ frequency.

The CTL signal is generated from the $\frac{1}{2} V_D$ signal so that both have the same frequency of 25 Hz (period 40 ms) when using the PAL video standard. Both waveforms are shown in Figure 2.12a. It can be seen from this figure that the duty ratio is 50 per cent for the $\frac{1}{2} V_D$ but not for the CTL signal. This is because the $\frac{1}{2} V_D$ signal is recorded with the video signal and is used for servo lock in the REC mode. It is used for tracking in the PB mode. (Remember that the PB signal is a string of positive and negative tip pulses if a square wave is recorded.)

It is difficult to identify the field sequence from the high and low levels of a square wave so the duty cycle of the $\frac{1}{2} V_D$ has to be changed before recording it. This is why the $\frac{1}{2} V_D$ is converted to the CTL signal. In Figure 2.12a the CTL timing edge is located at the trailing edge of a narrower pulse and appears in the even field of the video signal. The CTL signal indicates:

(a) the distribution of the recorded video tracks on the tape, i.e. a recorded period of the CTL signal corresponds to two tracks of the video signal on the tape;

(b) the field sequence, i.e. the narrow pulse width of a CTL signal is always recorded with the even field of a video signal; and

(c) the record phase, i.e. the time difference between recording both the trailing edge of the narrow pulse of the CTL track, and the leading edge of the synchronization of the even field on the video track, is always constant.

Therefore, if the tape transport speed in the PB mode can be maintained at the same speed as that in the REC mode, the PB CTL signal can be used as a reference signal to lock the servo system in the PB mode.

Because the fixed CTL (R/P) head is located approximately 74 mm from the tape-out point of the drum, the CTL trace and video trace being recorded simultaneously are a fixed distance apart on the tape. If the track pitch is

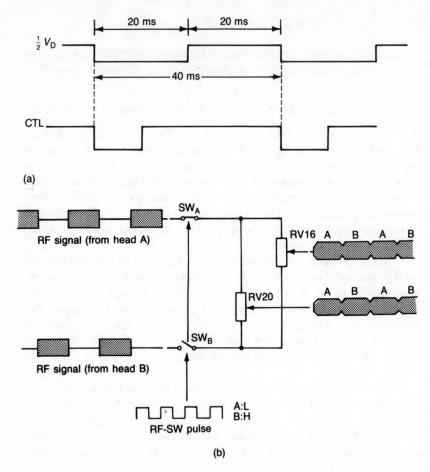

Figure 2.12 (a) Waveforms of the $\frac{1}{2} V_D$ signal and the CTL signal. (b) The RF signal picked up from the heads.

1.906 mm, the distance between simultaneously recorded video and CTL tracks is $74/1.906 = 38.824\,7637$ (or approximately 38) video traces, so the video trace is 19 periods of the CTL signal apart from the simultaneously recorded CTL trace.

The position of heads A and B

The switch pulse and the switch points of heads A and B are fixed on a rotating drum 180° apart and the drum takes 40 ms to rotate through 360°. Therefore, the signals picked up by the heads, known as the RF signals, have a 20 ms time difference as shown in Figure 2.12b. They are selected by electronic switches A and B and converged at balancing potentiometers RV20 and RV16.

Switches A and B are controlled by a switch pulse derived from the drum servo circuit.

The switch pulse, known as the RF-SW pulse in the U-matic format and the head-SW (H-SW) pulse in VHS, has a period of 40 ms and a pulse duty ratio of 50 per cent. The rising and trailing edges of the RF-SW pulses are timed to occur at two switch points. The point at which the rotary video head just comes into contact with the tape is called the IN-tape point while the point at which the video head separates from the tape is called the OUT-tape point. These two points are further described as IN-tape A and OUT-tape A for video head A, with similar terms for head B. Both IN-tape and OUT-tape points are collectively known as the switch points.

The distribution of magnets around the drums

In addition, a round plate is fixed between the upper and lower drums; it rotates with the upper drum and has a number of small detector magnets mounted on it. These magnets fall into two groups:

1. Eight magnets evenly distributed, i.e. magnets are fixed at 45° around a circle on the back of the plate with a radius r_1, for detecting the rotational speed of the drum.
2. A magnet is attached on a circle, radius r_2, on the back of the plate and is used to detect the rotational phase of the drum. This arrangement is shown in Figure 2.13b.

Two groups of detection coils are fixed on the upper side of the lower drum (see Figure 2.13a).

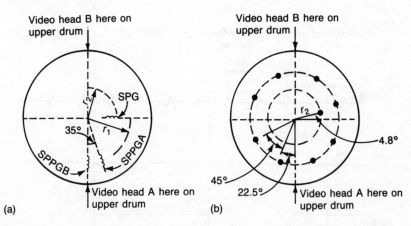

Figure 2.13   Location of the detector magnets on the drum: (a) shows upper side of lower drum and its relation to video heads on upper drum, (b) nine magnets are fixed on the rear of the plate.

Two coils, known as SPPGA and SPPGB (SPPG is a speed pulse generator), corresponding to the previously mentioned eight magnets, are located at intervals of 35° along the circle with radius r_1. One coil, known as the SPG (i.e. servo pulse generator) corresponds to the single magnet located on the circle of radius r_2. Using the IN-tape point A as a reference for locating the magnets on the plate which rotates with the upper drum, and coils on the lower drum, the coil of SPPGB is placed at an angle of 0° with video head A and the coil of SPPGA is placed in advance of SPPGB at an angle of 35°. The coil of SPG on the lower drum is advanced by 90° with respect to video head A on the upper drum.

On the stationary lower drum eight magnets, used to detect speed, are evenly distributed along the symmetrical axis passing through heads A and B on the upper drum, located at IN-tape A. The single magnet, used to detect phase, is placed at an advanced angle of 4.8° with respect to the coil of SPG as the drum is positioned at IN-tape A. According to this arrangement the magnets will pass the corresponding coil and generate a series of pulses from the coil. The pulses generated from coils SPPGA, SPPGB and SPG are used as timing pulses. Figure 2.14 shows the timing relationship of these pulses and the RF-SW pulse.

2.4.2 The drum phase servo system

The requirements of the drum servo system are that:

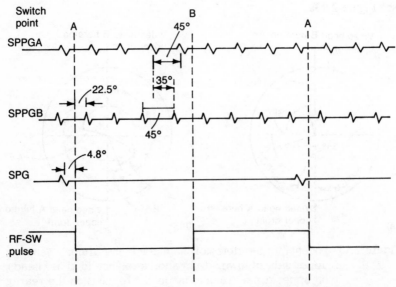

Figure 2.14 Timing relationship of SPPGA, SPPGB and SPG pulses.

1. During a recording period, the relationship between the position of the scanning video head on the tape and the video signal sent to the video head must follow a predetermined relationship. This is known as the recording phase.
2. During a replay period, the relationship between the PB signal from the video head and the position of the tracking video head on the tape must correspond with that during the recording period. This is known as tracking.

For a helical scanning VCR with two video heads, for example the VO-5850P, the two heads, i.e. heads A and B, have an angle of 180° between them and each video track scanned by a video head contains a field of the video signal. This is referred to as a non-segment scanning mode. Thus, two video heads on the drum are scanning two video tracks, respectively, so containing a frame of the video signal each time the drum rotates one complete revolution.

The following two schemes are used to deal with the discontinuity of the recording video signal as the switch between heads A and B takes place.

1. The angle in which the tape is in contact round the drum is a little more than 180°. This corresponds to a recordable length of about 6H of the video signal. It is therefore possible for both heads to be in contact with the tape simultaneously, i.e. the information of the 6H at the end of the front track is the same as that at the beginning of the next track. Of course, information will not be lost if the switch from the front track to the next track is chosen to be at the end of the 6H of the track as shown in Figure 2.15.

2. To prevent the effect of discontinuity at the switch point, caused by stretching of the tape length between the REC and PB modes, from being displayed on the screen, it is better to arrange the switch point at a field blank

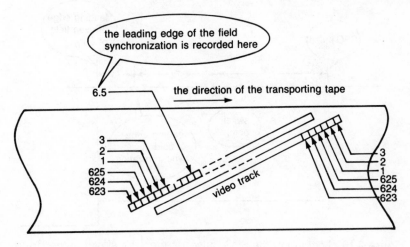

Figure 2.15 Switch from track to track at 6H.

period. For example, the switch point is arranged at a point 6.5H before the leading edge of the field synchronization in the U-matic format and those VCRs using a $\frac{1}{2}$-inch tape. The switch point is altered from 6.5 to 2.5H in machines using the high band U-matic format, i.e. U-matic(H). The switch point is maintained by the drum servo system.

For example, in the VO-5850P the SPG pulse, which indicates the rotational phase of the drum motor, is in advance of the switch point by 4.8° which corresponds to a time of:

$$\tau = \frac{40 \text{ ms}}{360°} \times 4.8° = 0.533 \text{ ms}$$

The $\frac{1}{2} V_\mathrm{D}$ signal, which indicates the reference signal, lags a constant time (τ) with respect to the leading edge of the field synchronization. Therefore, the question as to whether the switch point should be located at a point 6.5H before the leading edge of the field synchronization is answered by the fact that the SPG pulse is locked by the $\frac{1}{2} V_\mathrm{D}$ signal in the drum servo loop as shown in Figure 2.16.

2.4.3 The capstan phase servo system

The track format recorded on the tape is comprised of three parameters:

(a)   the recording phase;
(b)   the track pitch (P); and
(c)   the track angle ϕ.

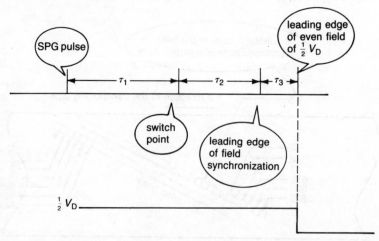

Figure 2.16 Locking of the SPG pulse by the $\frac{1}{2} V_\mathrm{D}$ signal in the drum servo loop; this ensures that $\tau_2 = 6.5H$.

The recording phase indicates that the information of each line in a field should be recorded in the pre-assigned place on the corresponding track. This can be achieved if it is ensured that:

1. The leading edge of the field synchronization is recorded at a point 6.5H after the switch point on the video track.
2. The rotating speed of the drum per revolution is uniform.

Obviously these points are the task of the drum servo.

A constant track pitch (P) is maintained by the capstan phase servo in the following manner. Because a frame of the video signal corresponds to two video tracks, the tape speed (v_t) is the same as the two track pitch $(2P)$ divided by the period of a frame T_v:

$$v_t = \frac{2a}{T_v}$$

or

$$v_t = 2f_v a$$

where f_v is the frame frequency, or 25 Hz for the PAL system. Therefore, locking the frequency generator (FG) signal, which is the sampling signal of the rotating speed of the capstan motor, and the $\frac{1}{2} V_D$ signal, ensures that the tape speed (v_t) varies directly with the frame frequency of the video signal (f_v):

$$v_t = Kf_v = \frac{Kv_t}{2a}$$

or

$$a = \frac{1}{2} K$$

where K is a proportional constant dependent on mechanical parameters such as the diameter of the capstan shaft, etc.

In the VO-2860P (a predecessor of VO-5850P), an FG signal is generated by rotary polarized permanent magnets, while in the VO-5850P a rotary polarized permanent magnet and a pair of Hall effect diodes are used to generate two FG signals, FG1 and FG2. When the capstan rotates at the normal speed, two signals are generated at 450 Hz with a phase difference of 90°.

The time difference between FG1 and FG2 indicates the rotational speed of the capstan motor. The direction of rotation of the capstan motor can be detected using a D-trigger as shown in Figure 2.17.

So to sum up, a constant track pitch (a) will be maintained if the FG signal from the rotating capstan motor is locked by the $\frac{1}{2} V_D$ signal in the capstan phase loop. The track angle is mainly dependent on the angle of the guide slot on the lower drum and the angle between the threading ring and the lower drum, i.e. dependent on the head-to-tape scanning mechanism.

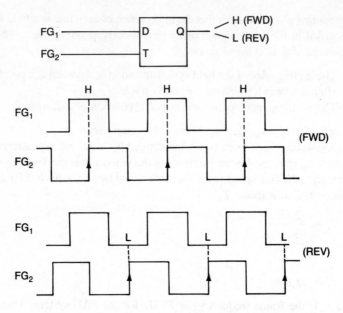

Figure 2.17 Detection of rotational direction of the capstan motor.
Note: the waveforms of FG1 and FG2 have been shaped.

2.4.4 Tracking in the PB mode

In the PB mode the first step is to lock the PB CTL signal and the $\frac{1}{2}V_D$ signal, so the tape speed is maintained to be the same as that in the REC mode. This is achieved by the capstan phase servo. Although the lock has been achieved, it is usually difficult to align the centre of the head gap and the centre of the video track. This is because:

1. When the player machine is not the same machine as the recorder, track errors such as the height of the video head on the drum, the distance between the CTL head and the switch point, exist between the two machines.
2. Even if the recording and playback machines are the same, different conditions during the period between recording and replay, such as temperature, humidity and tape stretch, exist.

To correct these errors the phase (or the delay) of the $\frac{1}{2}V_D$ signal should be adjusted to control the speed of the capstan motor and thus ensure tracking. The TRACKING button for controlling the tracking is on the front panel of the machine.

After the problems of the PB tape speed are corrected the PB CTL signal can be used as a reference signal to lock the rotating speed and phase of the drum motor. When the PB CTL signal is locked with the $\frac{1}{2} V_D$ signal, the $\frac{1}{2} V_D$ signal can also be used to indicate the rotational speed and phase of the drum motor in the REC mode and used to lock the SPG pulse to ensure tracking of the rotating drum, i.e. ensure that the drum is just rotating at the switch point when the video head picks up the information at a point 6.5H before the leading edge of the field synchronization.

2.4.5 The speed servo loop

With the exception of its phase servo the speed servo can be considered to be an auxiliary servo. It is used to suppress interference, and improve starting performance and smooth running of the motor. The following schemes are usually adopted in the speed servo.

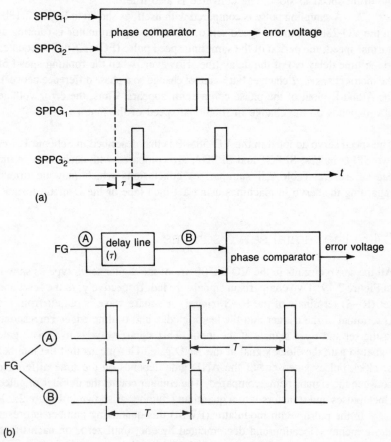

Figure 2.18 Drum and capstan speed servos: (a) drum speed servo, (b) capstan speed servo.

1. Two sampling pulses are compared as shown in Figure 2.18a. For example, in the drum speed servo of the VO-2860P, the two fixed SPPG coils are located at an angle of 35° with respect to each other so the time difference (Δt) which exists while the magnet rotates through the two coils, SPPGA and SPPGB, varies inversely with the rotating speed of the motor (n):

$$t = \frac{35°}{n \times 360°}$$

The output voltage from the phase comparator depends on the time difference between SPPG1 and SPPG2, i.e. depends on the rotating speed of the drum motor (n). This output voltage, referred to as the error voltage, is used to control the drum motor. When the rotating speed of the drum motor increases, the time difference (Δt) decreases and the error voltage also decreases forcing the drum speed to slow. The converse is also true.

2. A sampling pulse is compared with itself, as shown in Figure 2.18b. In the VO-2860P capstan speed servo, when the capstan motor is rotating at normal speed, the period of the sampling speed pulse (FG) is T which equates to the time delay (τ) of the delay line. However, when the rotating speed of the motor changes, T changes but τ cannot change so a phase difference between the A and B input of the phase comparator appears. Thus, the error voltage also depends on the change in rotational speed of the motor.

The speed servo adopted in the VO-5850P is that described in Scheme 1, i.e. two SPPG pulses, SPPG1 and SPPG2, are compared with each other in the capstan servo. Table 2.3 summarizes conclusions which may be drawn regarding the servo in machines using a $\frac{1}{2}$-inch tape or the U-matic format.

2.4.6 The digital servo scheme

All the servo circuits in the VO-5850P are of the digital servo type as shown in Figure 2.19. Two comparison signals are fed, respectively, to the reset and set (R−S) terminals of the R−S trigger. A square wave is output from the Q terminal of the trigger and the leading edge and trailing edges correspond to the set and reset signals at the input. This square wave is used as a gate signal to gate the clock signal at the AND gate. This means that the number of clock pulses gated out of the AND gate depends on the time difference between the signals being compared. The counter counts the number of gated clock pulses and transfers it to a quantizing number as shown in Figure 2.20.

In the pulse width modulator (PWM) this quantizing number is preset into a memory location and decremented by one, until zero, on each timer pulse. The output of the PWM is at a high logic level during the period between the presetting of the quantizing number and decrementing it to zero, and is

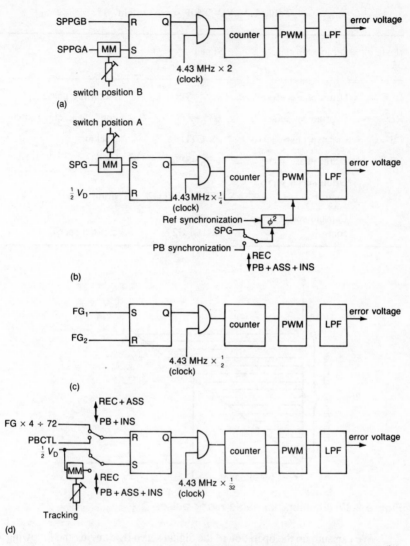

Figure 2.19 Digital servos in the VO-5850P: (a) drum speed servo, (b) drum phase servo, (c) capstan speed servo, (d) capstan phase servo.

at a low logic level outside this period. The pulse width of the square wave from the PWM is therefore varied directly with the quantizing number, and thus with the time difference between the two comparison signals, and hence motor speed. This square wave becomes a d.c. level at the output of the low pass filter and so the d.c. level, i.e. error voltage, varies with the pulse width of the square wave.

Table 2.3 VO-5850P servo system

Mode	Servo loop type	Signal comparison	Purpose
REC	Drum phase servo	$\frac{1}{2}V_D$–SPG	recording phase
REC	Capstan phase servo	$\frac{1}{2}V_D$–FG	constant tracking pitch
PB	Capstan phase servo	PB CTL–$\frac{1}{2}V_D$	tracking
PB	Drum phase servo	$\frac{1}{2}V_D$–SPG	tracking
	Drum speed servo control	SPPG1–SPPG2	drum speed
	Capstan speed servo control	FG1–FG2	capstan speed

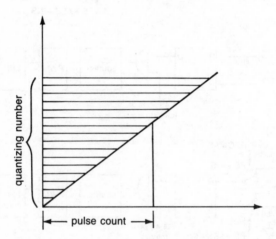

Figure 2.20 Counting of gated clock pulses.

We can sum up the operation of the digital servo by drawing the following conclusions.

If the motor is running slowly, the time difference between two comparison signals increases and the pulse width of the square wave from the PWM increases. This means that the error voltage which drives the motor is higher and the motor speed is increased. Transfer of the quantizing number to the PWM and decrementing it to zero takes a finite time through the tristate buffer provided for this purpose, so the circuits involved are controlled by a reference clock signal. When a high logic level of the reference clock signal is present, the switch turns on and the buffer is turned into its high impedance state.

Transfer of the data to the PWM and decrementing to zero are carried out while the buffer is in this state. Conversely, a low logic level of the reference clock signal turns the switch off and the tri-state gate on. The error voltage is then applied to the motor. Buffering and temporary storage of the data in this way avoids incomplete transfer of data.

Correct choice of the clock signal frequency ensures that the counter data is located at the centre of the oblique line as shown in Figure 2.20. Since the required change in speed of the motor is large in the search mode, the slope of the oblique line and the frequency of the timer pulse also need to be altered. Operation in this case is the same as before except that the dynamic range has been changed.

Since the periods of the comparison signals as shown in Figure 2.19 are different, the frequency of the clock signal is also different. For example, the frequency of the SPPG pulse is eight times greater than that of the SPG pulse, so the clock frequency in the latter case is eight times greater than that required in the case of the SPPG pulse.

2.5 Basic edit concepts

1. Usually there are two kinds of editing mode. These are the insert editing mode (INS) and the assemble editing mode (ASS). The ASS mode is mainly used to connect prior recorded material with material being currently recorded as shown in Figure 2.21a. It has an 'edit-in' point only and all signals (such as video, audio and CTL signals) are recorded simultaneously. The INS mode is used to insert a section of new signal (video or audio signals) on a prerecorded signal as shown in Figure 2.21b. Thus this mode needs two edit points (edit-in and edit-out) and may insert any signal (video or audio), or may insert both video and audio together, but it is necessary to keep the previously recorded CTL signal.

2. It is necessary to use rotary erase heads instead of the stationary full erase head during editing. Using stationary erase heads would leave triangular overlap and blank areas of the prerecorded sections of tape during edit switching, as shown in Figure 2.22. There are two rotary erase (RE) heads fixed on the head drum as shown in Figure 2.23. These are referred to as RE_A and RE_B. RE_A is placed at $32°$ before the corresponding video head A. Similar mounting arrangements are used for RE_B and video B. To ensure overlap between the locus of RE and the video traces, it is necessary to place RE rather higher than its corresponding video head. The height of RE above the video head (h) can be calculated as follows, referring to Figure 2.24:

$$h = \frac{32°}{180°} H = \frac{32°}{180°} a \tan \theta$$

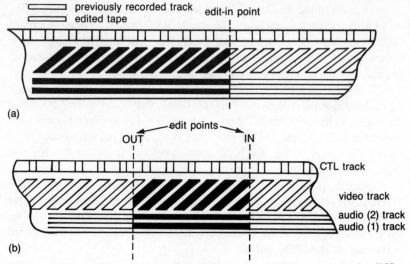

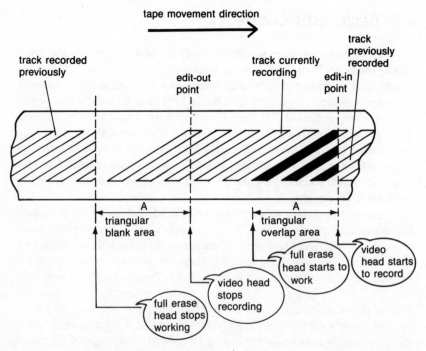

Figure 2.21 (a) The track in ASS edit mode. (b) The track in the INS edit mode.

Figure 2.22 Use of rotary erase heads during editing. Note: (a) it is in INS mode; (b) *A* is the distance between the full erase head and the point where the video head starts to rotate into the tape.

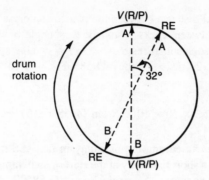

Figure 2.23 Location of rotary erase heads on the drum.

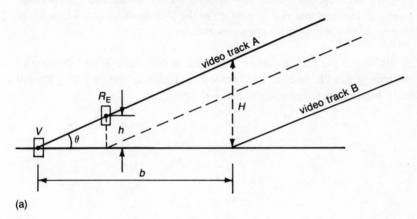

(a)

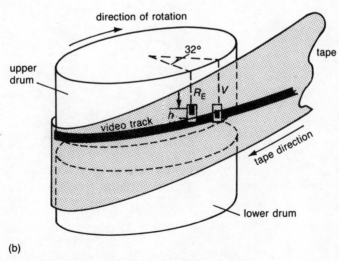

(b)

Figure 2.24 Height of rotary erase heads on the drum.

where h is the vertical distance between the record and erase heads; H is the vertical separation between tracks A and B; θ = angle of the video track; b = interval of two adjacent tracks $a = v_t\, T_v$; v_t = tape speed; and T_v = field period of video (20 ms). For the VO-5850P $\theta = 4°58'6.2''$ and v_t = 9.53 cm/s, therefore

$$h = \frac{32}{180} \times 9.53 \times 20 \times 10^{-3} \times \tan\, (4°58'6.2'') = 0.03 \text{ mm}$$

Because there is an editing-in point only in the ASS mode, the RE is needed to operate for a short time only at the start of switching from PB mode to REC mode. This is as short as 5 s on the Sony VO-5850P. After this period the full erase head can operate instead of the RE. However, there are two edit points, at the start and end of editing, in the INS mode so that RE needs to operate during the whole editing process.

In addition to the editing modes mentioned above, there is also the case of switching from PB mode to REC mode to consider. In this case only the full erase head is operating and the RE is disabled.

Chapter 3

The video channel

The main circuits in the $\frac{3}{4}$-inch tape VCR (including U-matic and U-matic(H)), $\frac{1}{2}$-inch tape VCR (including VHS and Betamax, etc.) or 8 mm tape systems, consist of video and audio channels, servo systems and a control system. In addition, some optional circuits, such as a tuner and televison demodulator, RF converter and circuits displaying mode and time, are included in some machines, particularly the domestic VCR.

Because the audio channel is the same as that in an audio recorder, only the above circuits will be described in the following chapters, but high fidelity audio techniques will be discussed as a development of modern VCRs. For clarity, circuit analysis will be described first in terms of machines with U-matic format, followed by examples of the circuit used in the VHS system.

The video channel of a VCR of $\frac{3}{4}$-inch or 8 mm tape consists of a record (REC) and a playback (PB) channel, while each channel is further divided into two paths. These are the luminance (Y) and chrominance (C) paths. The REC Y, REC C, PB Y and PB C channels will be discussed first in relation to a U-matic format VCR, followed by examples in the VHS format.

3.1 REC Y channel

As described earlier, the nucleus of the REC Y channel is the frequency modulator (FM). However, the video (Y) signal needs processing before modulation. There are two forms of video signal processing to ensure satisfactory operation of the modulation and to improve the signal-to-noise (S/N) ratio of the video signal during modulation.

3.1.1 Video signal processing

The video signal is processed in the circuits shown in Figure 3.1 before

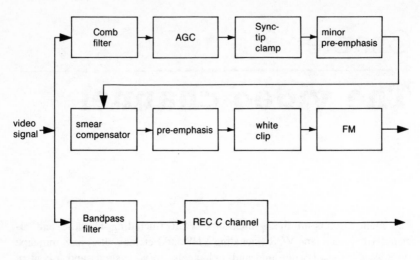

Figure 3.1 Video signal processing schematic.

frequency modulation. The auto-gain control (AGC) and sync-tip clamp circuits ensure satisfactory modulation, while the pre-emphasis, white tip clip, smear compensator, minor pre-emphasis and comb filter work to improve the S/N ratio of the video signal. The operation principles of these circuits are discussed below.

AGC loop

The AGC loop consists mainly of an AGC amplifier, synchronizing separator (sync-separation), Y and synchronizing signals mixer, peak value detector and d.c. amplifier. These are shown in Figure 3.2. The AGC amplifier features a d.c. voltage gain control. The gain of the amplifier decreases with increasing $V_{d.c.}$.

The video signal output from the AGC amplifier is fed to the sync-tip clamp and the sync-separator. The synchronizing signal is separated and delayed slightly, then it is mixed with the video output from the sync-tip clamp. The H-sync pulse output from the sync-separator is also passed to the sync-tip clamp to be used as a clamping pulse, and to the chroma path to be used to control the burst gate. The mixed signal is fed to the peak value detector to provide a d.c. voltage which is relative to the level of the input video signal, and results in a synchronizing pulse amplitude that is related to the video signal amplitude. The advantage of mixing the synchronizing pulse is that the d.c. voltage controlling the gain of the AGC amplifier does not vary with the changing video picture contents.

The d.c. voltage is divided using fixed and variable (RV) resistors to control the gain of the AGC amplifier and change the frequency deviation of

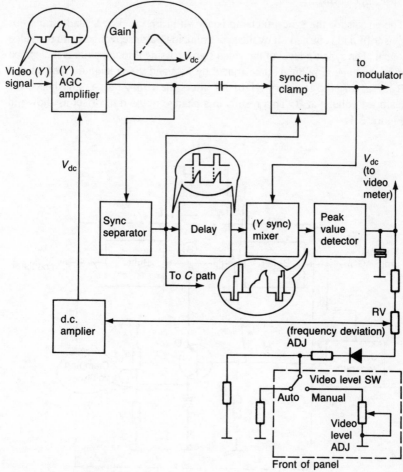

Figure 3.2   The AGC loop.

the frequency modulator. The variable resistor is called the frequency deviation adjust (ADJ). The video level switch (SW) and ADJ are available on the front panel to permit adjustments when the condition of the input video signal is higher or lower than specification. In this case it is necessary to set SW to the 'manual' side and then adjust RV up or down to change the d.c. voltage, and thus the gain of the AGC amplifier, to ensure correct operation of the frequency modulator.

Normally, SW must be set to the 'auto' side and the ADJ to the midposition. Because the d.c. voltage output of the peak value detector represents the level of the input video signal, it is measured using a d.c. voltage meter on the front panel to indicate the video signal level.

Sync-tip clamp

The circuit for the sync-tip clamp is shown in Figure 3.3. Usually transistor Q is 'off' and is turned on by the synchronizing pulse separated from the video signal. The sync-tip level of the video signal passing through point B is clamped to the voltage at point A, determined by V_{CC} and the voltage divider R1 and R2. Varying the resistors R1 or R2 alters the voltage at A, and hence the clamped voltage at B. This results in a change in the d.c. offset as shown in Figure 2.1.

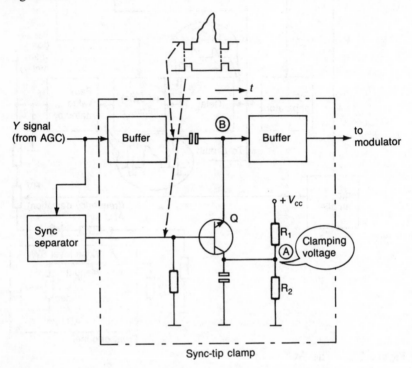

Figure 3.3   The sync-tip clamp.

Comb filter

Figure 3.4 shows how the adoption of low pass filtering in the REC Y channel suppresses the C signal from the video signal. Most VCRs have adopted this method of suppression. It is difficult to choose the cut-off frequency of the LPF since, if the cut-off frequency is too high the interference from the C signal will be aggravated, and if the cut-off frequency is too low, the resolution of the PB picture will be reduced. A comb filter is used to prevent this. The principle diagram of the comb filter is shown in Figure 3.5a, and Figure 3.5b shows its transfer characteristic.

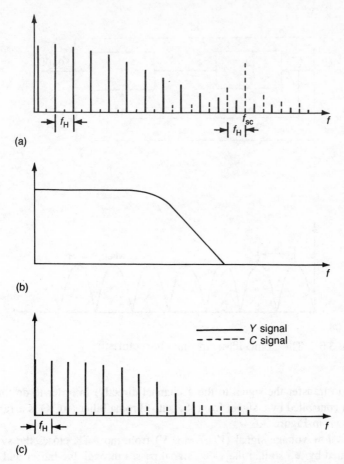

(a)

(b)

(c)

Figure 3.4 Filtering of the video signal: (a) frequency spectrum of the video signal, (b) frequency response of LPF, (c) frequency spectrum of the signal passed through LPF.

If the delay time τ is chosen to be equal to H, the period of the H-sync pulse, the interval between two wave peaks is the horizontal frequency f_H. Because the energy of the video signal is distributed within the pass band of the comb filter at intervals of f_H the luminance signal (Y) is passed and the chrominance signal (C) is attenuated.

Although the frequency response is satisfactory, the response of the filter could deteriorate the phase characteristic of the video signal passing through the LPF, or comb filter, so it is necessary to arrange for phase equalization following the filter. A phase equalizer is a network which could correct the phase distortion and maintain the original frequency characteristic.

If the input video signal is black and white only, i.e. no C signal, it is

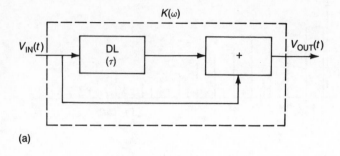

(a)

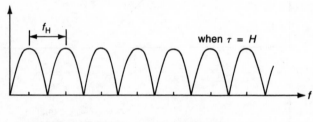

(b)

Figure 3.5 The comb filter and its characteristic.

better to transfer the signal to the Y channel directly. In order to do this, a switch controlled by a signal from the auto chroma killer (ACK) is arranged as shown in Figure 3.6.

A low voltage signal (V(L) = 0 V) from the ACK closes the switch (indicated by ●) so that the video signal passes through the buffer and into the Y channel. If the video signal is a colour signal a higher voltage (V(H)), say +5 V, is sent by the ACK to open the switch (indicated by ○). This directs the video signal through the comb filter and phase equalizer before transferring to the Y channel.

Pre-emphasis circuit

Frequency deviation is generally limited to a specific value in an FM system, so the modulation factor decreases as the frequency of the modulating signal increases. This lowers the S/N ratio in the high frequency region after the FM signal is demodulated. In order to improve this, the pre-emphasis circuit is arranged before modulating in the REC channel and the de-emphasis is placed after demodulation in the PB channel.

In order to maintain the original frequency characteristic during S/N improvement it is necessary that both circuits have the same time constant. Two types of pre-emphasis networks are shown in Figure 3.7.

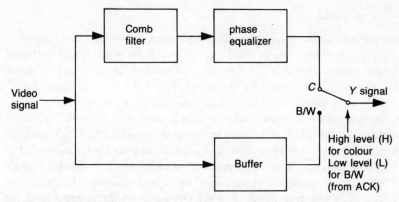

Figure 3.6 Auto chroma killer switching of the comb filter.

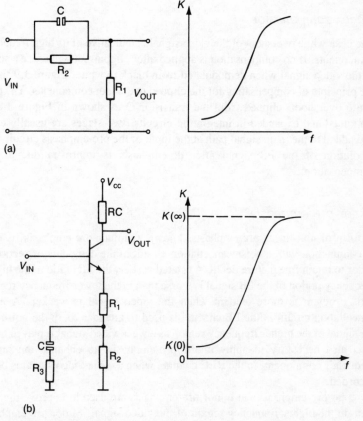

Figure 3.7 Pre-emphasis networks: (a) pre-emphasis network without source, (b) pre-emphasis network with source.

White clipper

The purpose of arranging the white clipper after the pre-emphasis circuit is to prevent over-deviation of the frequency modulator which would be caused by peak white overshoot of the video signal. The white clipper is used to remove this overshoot. The circuit for the white clipper is shown in Figure 3.8.

Usually the transistor Q is off and the video signal is passed unaffected. As an overshoot level of peak white arrives at point A, it is coupled to ground by the capacitance C1 via the base-emitter of the transistor Q, which has been turned on by the overshoot level. The clipped overshoot level is directed to the smear compensator from the collector of transistor Q. A variable resistor (RV) is used to adjust the clipped level of white overshoot. The diode (D), resistor (R3) and capacitance (C2) are chosen to set the clipping level for overshoot in the sync-tip. Usually the diode is cut off and turned on by the overshoot in the sync-tip which is then coupled to ground by capacitance, C2.

Smear compensator

The peak white overshoot of the video signal is indicative of its high frequency components. If no compensation is applied after clipping, it will cause a smear of the video signal when demodulated from black to white. Figure 3.9 shows the principle of compensating for the clipped overshoot components. The peak white overshoot, clipped from the transistor Q as shown in Figure 3.8, is smoothed and expanded in integrating circuits (two stages are usually used) and added to the main signal path at the input of the pre-emphasis circuit. The resolution of the PB signal after de-emphasis is improved due to this compensation.

Minor pre-emphasis

A form of non-linear pre-emphasis, known as minor pre-emphasis, is used in conjunction with a noise cancellation circuit in the demodulator section in order to retain fine picture detail. As stated earlier, the S/N ratio in the higher frequency region of the PB signal is worse than in the lower frequency region. This problem is more evident when the video signal is weaker. A noise cancellation circuit in the PB channel is used to attenuate both the noise and the signal in the higher frequency region as a weak video signal is played back. It is also necessary to apply minor pre-emphasis, to enhance the higher frequency component, in the REC channel when a weak video signal is being recorded.

So pre-emphasis and minor pre-emphasis are used to improve the S/N ratio in the higher frequency region of the video signal. Minor pre-emphasis is used under weak signal conditions while pre-emphasis is used under both weak and normal video signal conditions. The minor pre-emphasis circuit is

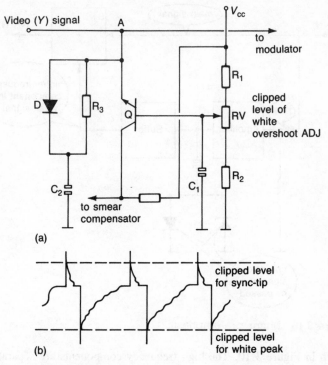

Video (Y) signal

A

V_{cc}

to modulator

R_1

clipped level of white overshoot ADJ

RV

D

R_3

Q

R_2

C_2

C_1

to smear compensator

(a)

clipped level for sync-tip

clipped level for white peak

(b)

Figure 3.8 The white clipper: (a) clipping circuit for white peak and sync-tip, (b) video waveform.

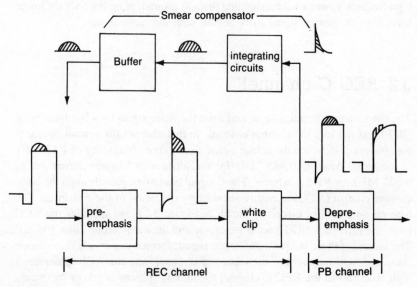

Smear compensator

Buffer

integrating circuits

pre-emphasis

white clip

De-emphasis

REC channel

PB channel

Figure 3.9 Clipped overshoot compensation.

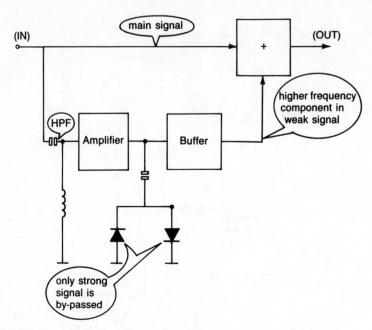

Figure 3.10 Minor pre-emphasis.

shown in Figure 3.10. The high frequency components are separated from the video signal by an LC network used as a high pass filter (HPF) and then amplified. Two diodes between the amplifier and a buffer clip the higher level, high frequency components, shunting them to ground, allowing only the lower level high frequency signal to be mixed with the main signal.

3.2 REC *C* channel

The chrominance signal is separated from the video signal by a bandpass filter (BPF) and fed into the chroma-channel. In this channel the central device is the converter, in which the colour signal's sub-carrier frequency of 4.43 MHz is converted down to 0.685 MHz for machines with U-matic format, or to 0.627 MHz for VHS machines. The *C* signal has to first pass through the auto-chroma control (ACC) to ensure satisfactory operation of the converter. The auto-chroma control circuit is shown in Figure 3.11 and includes the ACC loop, the reference (REF) burst generator and the auto-colour killer (ACK). The second of these feeds the reference signal, formed to provide information about the field numbers of the input video signal, and the ACK generates a high level (H) to the REC *C* channel for muting if there is no, or too weak, colour signal in the input video signal.

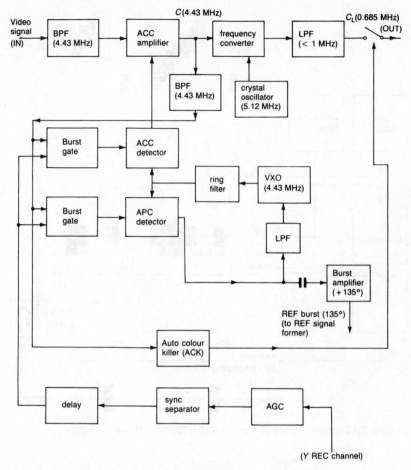

Figure 3.11 The auto chroma control circuit schematic.

Comparison between sampling burst and a standard oscillating signal is used in both the ACC loop and the REF burst generator. The sampling burst is separated from the chroma signal by the burst gate in which the delayed sync-signal from the AGC in the REC Y channel is used as a gate control as shown in Figure 3.12. This samples the amplitude and phase of the input chroma signal separately in the ACC loop and REF burst generator. The standard oscillating signal needs to be locked in phase with the input chroma signal. The phase locking principle is shown in Figure 3.13. The VXO consists of an IC operational amplifier with a coupling transformer (T1) providing the feedback loop. The frequency of oscillation is determined by the crystal (X) and the variable capacitor controlled by the d.c. voltage (D) from the LPF. This oscillating signal is fed to the APC and ACC detectors via the ring filter. The burst and oscillator signals are compared in phase in the APC detector,

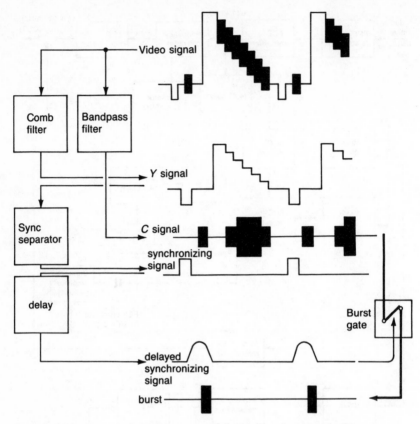

Figure 3.12 Sample burst separated from the chroma signal.

and in amplitude in the ACC detector. The output voltage of the APC detector, which includes components of a.c. and d.c., is determined by the difference in phase between the burst and oscillator signals.

After low pass filtering the d.c. component of the output voltage is used to control the capacitance of D, further locking the phase of the VXO. In the ACC detector the output (d.c.) voltage is proportional to the amplitude difference between burst and oscillating signals. Because the amplitude of the burst is one third of the amplitude of the chroma signal and is also larger than the oscillator signal, the output (d.c.) voltage increases as the input chroma signal increases. Similarly, the output (d.c.) voltage decreases as the input signal decreases. The gain of the ACC amplifier therefore decreases with increasing amplitude of the input chroma signal. This keeps the output of the amplifier generally constant to suit the requirements of the frequency converter. This is the same as the AGC loop in the REC Y channel.

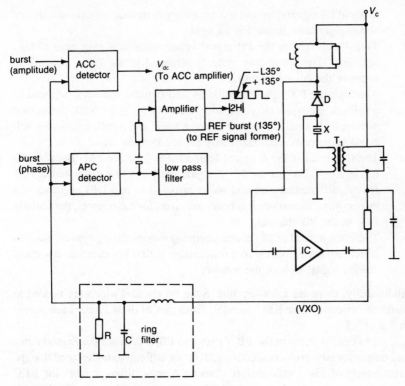

Figure 3.13 The phase locking principle.

3.3 PB *Y* channel

3.3.1 Outline

Because the PB signal originates from the head-to-tape scan, there are a number of special points regarding the difference between the PB signal/channel and the REC signal/channel which should be noted. These are:

1. The upper sideband of the modulated (or RF *Y* signal) is attenuated by the PB frequency characteristic of the video head which results in an imbalance in the upper and lower sideband amplitudes. It is necessary to arrange a sideband equalizer in the PB channel to compensate for this imbalance.

2. Drop-out of the PB signal can occur when the magnetic powder on the tape surface is damaged or removed, or is covered by dust. In order

to avoid PB signal drop-out it is necessary to include a drop-out detector and compensator in the PB channel.

3. Time-base error in the PB signal is unavoidable during head-to-tape scanning. Time-base correction is arranged in the PB C channel to improve the colour picture.

4. Although the PB signal from the A and B paths has been adjusted in amplitude by equalizing potentiometers RV16 and RV20, mentioned previously, a small difference between the A and B path amplitudes still needs to be corrected, especially for the chroma signal.

5. The C_L signal is the C signal (chrominance) before FM. Because the C_L signal 'rides' on the RF Y signal until they are separated during replay, differential gain and phase errors (DG and DP) occur caused by non-linear distortion. It is better to correct for these errors, particularly DG, in the PB channel.

6. There are a number of special operating modes during replay, such as search and pause modes, so it is necessary to develop an artificial vertical driving signal to lock the monitor.

Additionally, there are a few circuits in the PB channel which are needed to match the circuits in the REC channel. Examples of these circuits are shown in Table 3.1.

The central items in the PB Y path and PB C path are the demodulator and frequency converter, respectively. It is not difficult to understand that the arrangement of PB circuits before demodulation is different from the REC circuits. For example, there is no AGC circuit required in the PB channel. We can use the arrangement of circuitry in the PB channel of the Sony VO-5850P to help our understanding. This arrangement is shown in Figure 3.14.

The PB RF Y signal and PB C_L signals are separated from the PB pre-amplifier by a high pass filter (HPF) and low pass filter (LPF) respectively. The two PB-EE selectors are selected for either the PB mode, or non-PB modes such as REC, FF, REW and STOP.

Table 3.1 Equivalent circuits in PB and REC channels

In REC channel	In PB channel
Frequency modulator (FM)	Frequency demodulator (DM)
Pre-emphasis	De-emphasis
Minor pre-emphasis	Noise cancellation
Downconverted subcarrier	Upconverted subcarrier

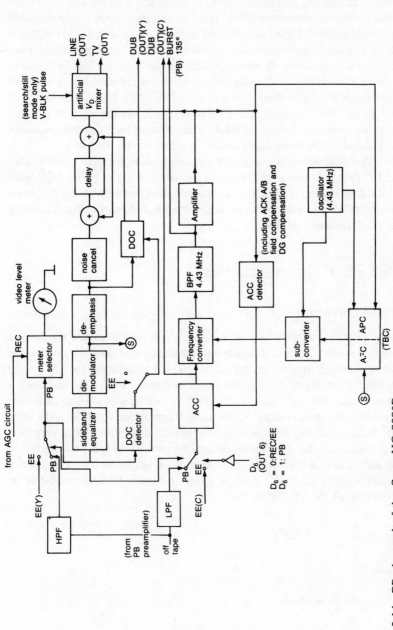

Figure 3.14 PB channel of the Sony VO-5850P.

The video channel **67**

In the PB Y channel, the RF Y signal, via a sideband equalizer, demodulator and de-emphasis circuit, becomes the Y signal and is then fed to the drop-out compensator (DOC) and noise cancellation circuit in which the demodulated Y signal is combined with the reconverted (4.43 MHz) C signal. The final stage features the introduction of a vertical blank pulse from the servo system during the special effects PB modes or the SEARCH mode. In the normal PB modes, the stage uses only the output amplifier. The output video signal from this stage is fed to the LINE-OUT BNC connector.

In the PB C channel, the C_L signal is fed, via the ACC circuit and frequency converter, to become a 4.43 MHz C signal and to be combined with the Y signal in the noise cancellation circuit of the PB Y channel.

There are also a few ancillary circuits, such as the auto chroma killer (ACK), A/B field memory circuit and DG compensation, in the ACC loop. A time-base corrector (TBC) is also used in conjunction with the frequency converter. We will discuss the circuits of the PB Y channel before discussing the PB C channel circuits.

3.3.2 PB Y channel circuits

Cosine corrector

Using a cosine corrector as a sideband equalizer to compensate for the loss in high frequency, results in an obvious advantage as it does not cause phase distortion at high frequencies. It is, therefore, not necessary to include a phase equalizer after improving the frequency response. Figure 3.15a shows a sideband equalizer and Figure 3.15b shows the principle of a cosine corrector. The corrector consists mainly of a differential operational amplifier and a delay line. It can be shown that because the operational amplifier has a very high input impedance, the signal passes through the network without phase distortion and suffers only a time delay (τ). The network has an amplitude/frequency characteristic of the cosine type, as shown in Figure 3.16. If $\tau = 50$ ns, as in the case of the VO-5850P then:

$$f_0 = \frac{1}{4\tau} = 5 \text{ MHz}$$

where

$$\omega_0 = \frac{\pi}{2\tau} = 2\pi f_0$$

So the output signal of the cosine corrector will increase when $f > 5$ MHz and will be attenuated when $f < 5$ Hz, resulting in the compensating characteristic of the cosine curve.

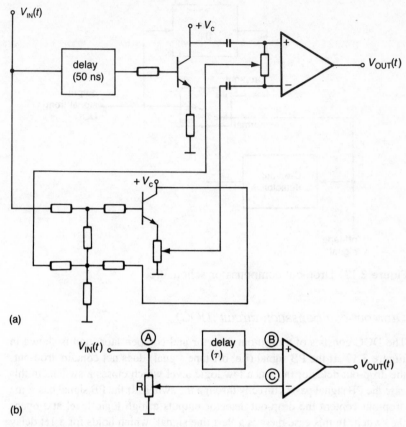

(a)

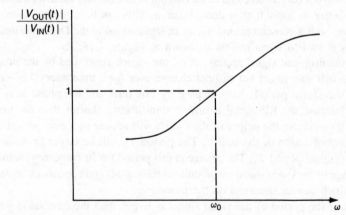

(b)

Figure 3.15 (a) Sideband equalizer. (b) Cosine corrector.

Figure 3.16 Amplitude/frequency characteristic of the cosine corrector.

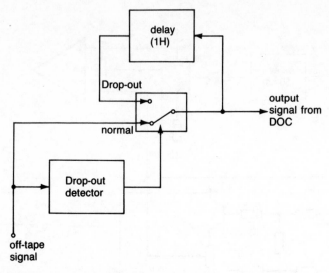

Figure 3.17 Drop-out compensator schematic.

Drop-out compensation circuit (DOC)

The DOC consists of a drop-out detector and compensator, and is shown in Figure 3.17. If the PB signal (i.e. off-tape signal) does not contain drop-out, the drop-out detector outputs a low logic level which closes a switch. In this case the PB signal passes directly through the switch. If the PB signal has some drop-out content the drop-out detector outputs a high logic level and opens the switch. In this case there is a 'last line signal' which holds for a 1H delay and remains on the drop-out terminal of the switch to replace the dropped out signal. Drop-out correction could be carried out before or after the demodulator, but it is better to apply it after demodulation. This can be explained using a sine wave, with a constant period T_1, as an input signal to the DOC (normally the input is an FM signal). This is shown in Figure 3.18.

If the drop-out signal appears at t_1, the switch controlled by the drop-out pulse will start to act but will not change over for a time interval $t_2 - t_1$. Several waveform periods have elapsed by this time and the phase at t_2 is random because the RF signal period is significantly shorter than the time interval $(t_2 - t_1)$. So the output of the switch will appear as a new period T_2 at the point of action of the switch. The period T_2 will be larger or smaller than the original period T_1. The change in this period (or its frequency) results in a change in the video signal (demodulated RF signal) and appears as 'switch noise' which can be observed on the monitor.

Since the period of the video signal is larger than the interval $t_2 - t_1$, the change in the video signal is very small during the switch-over. The switch noise is also very small if the DOC follows demodulation.

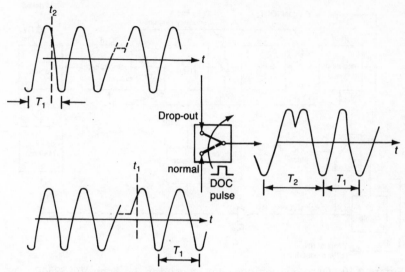

Figure 3.18 Drop-out correction.

The DOC in some VCRs of U-matic format, such as the VO-5850P, is placed after the demodulator but the predecessors of this model, VO-2860P and VHS, etc., had the DOC arranged before demodulation. Our discussions on DOC will concentrate on the VO-5850P as an example to explain the principles of DOC.

The DOC consists of two main functions; the drop-out detector and the drop-out compensator.

DROP-OUT COMPENSATOR

The block diagram of the drop-out compensator, as found in the VO-5850P is shown in Figure 3.19, and includes equalizing circuits and a delay for a period of 1H. Because the demodulated Y signal falls outside the response of the delay line it cannot be passed through a 1H delay and has to be modulated onto a carrier of 9.3 MHz. The Y signal, heterodyned with the 9.3 MHz carrier, is buffered and passed through a one-line ultrasonic delay. (The reason a 9.3 MHz oscillating frequency is chosen is to ensure that the frequency range of the heterodyne signal falls into the frequency response of the delay line, which is set by LV37, LV42 and RV19.) The heterodyning signal is then recovered as the video signal by a complementary balanced demodulator circuit, IC18.

To ensure that the compensating signal is as close as possible with the dropped out signal, in phase (or delay time), amplitude and d.c. level, delay time adjustment is provided at Q407 and DOC level is set by RV22.

The d.c. level balance circuit of the compensating signal is used to balance the d.c. level between the normal and 1H delayed signals during the H-sync

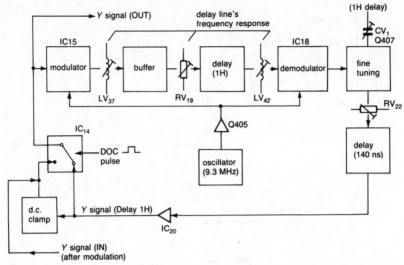

Figure 3.19 Drop-out compensator as used in the Sony VO-5850P.

period as shown in Figure 3.20. When the synchronizing pulse appears at TP8, the transistor Q603 turns on and the synchronizing pulse is of the delayed signals fed to the emitter of Q603. At the same time the synchronizing pulse of the normal signals is applied to the base of transistor Q601 and turns it on. This means that the divided voltage formed by RV61, R603 and the output impedance of Q601 will clamp the delayed signal. The adjustment of RV61 ensures that the d.c. level of the delayed signal is the same as that of the normal signal.

DROP-OUT DETECTOR
The block diagram of the drop-out detector in the VO-5850P is shown in Figure 3.21 and the waveforms at each part of this block diagram is shown in Figure 3.22.

The detector detects the drop out of the PB RF Y signal, shown as waveform 1, on which there is also high frequency (HF) noise and components of parasitic modulated amplitude in addition to the drop-out signal. The high frequency noise will be cancelled by the limiter (waveform 2) but the component of parasitic modulated amplitude and the drop-out signal still remain.

After the envelope detector, the envelope of the RF signal takes the waveform shown in 3. It is clipped and shaped by a Schmitt trigger into waveform 4, in which the parasitic components have been clipped by adjusting the threshold level (RV22). Waveform 4 is inverted by the inverting amplifier giving waveform 5a. The drop-out of the signal between two fields always appears during the alternate switch-over of the two heads and requires compensation. In order to do this, the RF SW pulse, which is shown as waveform 8, is used to generate the DOC pulse at the switching points. The

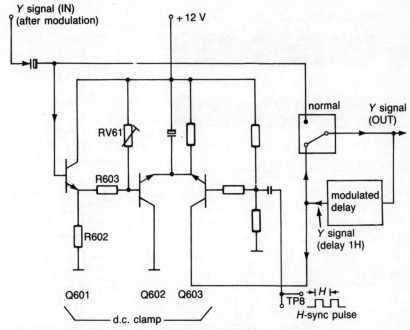

Figure 3.20 The d.c. level balancing circuit.

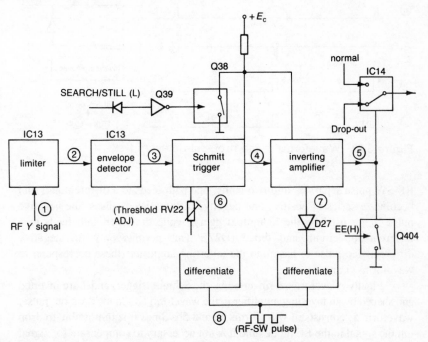

Figure 3.21 The drop-out detector in the VO-5850P.

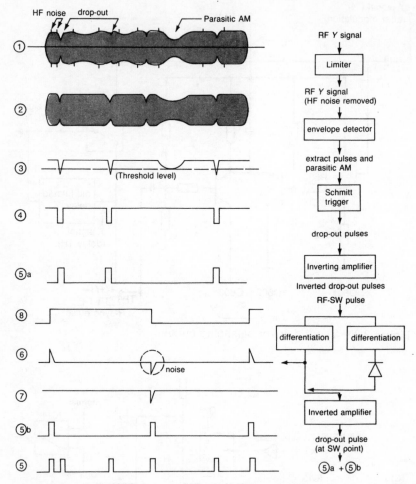

Figure 3.22 Waveforms in the drop-out detector.

RF-SW pulse separates into two paths; one path feeds to a differentiator and becomes a series of positive and negative differentiated pulses (the negative pulse is not used by the Schmitt trigger), while the other path feeds to a differentiator circuit and diode (D27). This permits only the negative differentiated pulse to pass into the inverting amplifier (these correspond to waveforms 6 and 7).

Finally, waveforms 6 (inverted by the Schmitt trigger) and 7 are inverted and shaped by an inverting amplifier to the waveform 5b. So the drop-out pulse, waveform 5, consists of waveforms 5a and 5b. Since it is impossible to drop out the signal in the EE mode, and it is not necessary to compensate for signal drop-out in the search mode, the DOC needs to be suppressed in these cases.

To do this, a low level, known as SEARCH/STILL(L), is applied to transistor Q38 through inverter Q39 and turns it on. This shorts the power supply, $+E_c$, of the Schmitt trigger and inverter. The control signal, SEARCH/STILL(L), is sent from the microcomputer control system.

Noise cancellation circuit

The noise cancelling circuit, as found in VCRs utilizing minor pre-emphasis, is shown in Figure 3.23. This circuit consists of a main path and a branch. In the branch path the Y signal, containing noise, passes through the inverter (Q25), bandpass filter (BPF) (C97 and L, C420, R612) and the linear amplifier (IC11).

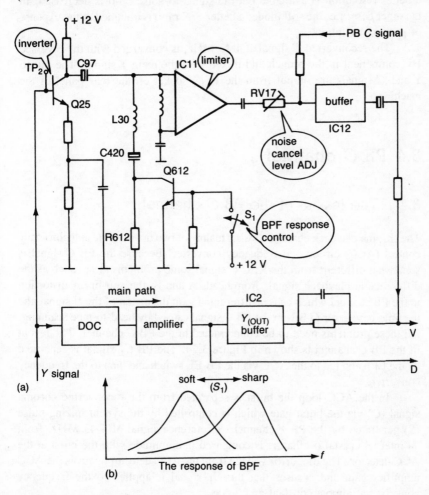

Figure 3.23 Noise cancellation.

The HF components, signal and noise, separated by the BPF, are level set by RV and then combined in opposite polarity with the Y signal from the main path. The HF noise components will be removed in this inverse mixing path and the S/N ratio will be improved. The HF signal components, which are lost, are compensated for in the minor pre-emphasis circuit in REC, described earlier. The resonance frequency of the BPF depends on the capacitances C97 and C420, the inductance L and resistor R612, so a change in R1 will move the resonance as shown in Figure 3.23b.

There is a manual mini-switch (S1) on the demodulator board (DM22) for turning on the transistor Q612 and shorting the resistor R612. In this case, i.e. the sharp mode, the resonance frequency is increased with the result that a better resolution is available but the signal-to-noise ratio is decreased. In the other case, i.e. the soft mode, a better S/N ratio is obtained at the expense of poorer resolution.

The reconverted C signal at 4.43 MHz, is converged with the reversed HF component in the branch and mixed with the main Y signal. The mixed Y and C signals are output from the V_{OUT} socket on the rear panel of the machine.

3.4 PB C channel

3.4.1 The layout of the PB C channel

The chroma playback channel consists mainly of two devices, the autochromatic control (ACC) circuit and frequency converter. Because the PB C signal is somewhat different from the REC C signal some problems arise, such as the difference in playback signals from heads A and B, the non-linear distortion in the PB channel which causes differential gain distortion of the C signal and the time-base error (TBE) in the PB C signal caused by head-to-tape scanning. All these problems need to be counteracted in the PB C channel. The layout of the PB C channel is shown in Figure 3.24. The PB C_L signal is separated by the LPF and fed to the ACC via the PB-EE switch and then to the frequency converter.

In the ACC loop the burst is separated from the reconverted chroma signal (C) via the burst gate which is controlled by the synchronizing pulse (S) generated by the PB Y channel. A reference signal of 4.43 MHz, from an internal crystal oscillator, is compared in amplitude with the burst at the ACC detector. The d.c. error voltage from the ACC detector controls the ACC amplifier gain and ensures that the C_L signal is applied to the frequency converter at almost constant amplitude.

In the frequency converter the synchronizing signal and burst signal both

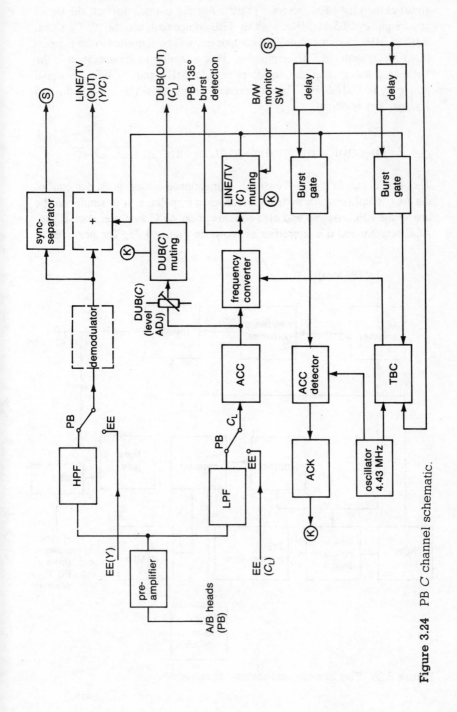

Figure 3.24 PB *C* channel schematic.

contain TBE so they are compared in frequency and phase with a reference signal at the time-base corrector (TBC). For the U-matic format, the signal at a frequency of 5.12 MHz, with its TBE component, and the PB C_L signal at 0.685 MHz, with the same TBE component, will be converted to the C signal (4.43 MHz) with the TBE eliminated. This is similar to the operation in the VHS case except that the signal is at 5.06 MHz and the PB C_L signal frequency is 0.627 MHz. The reconverted C signal mixes with the modulated Y signal and is output.

3.4.2 The autochromatic control circuit (ACC)

The ACC circuit in the PB C channel, as mentioned earlier, is used to control the ACC amplifier gain by using feedback d.c. voltage. It is similar to the one in the REC channel and also consists of an ACC amplifier, burst gate, ACC detector and d.c. amplifier as shown in Figure 3.25. The principle of

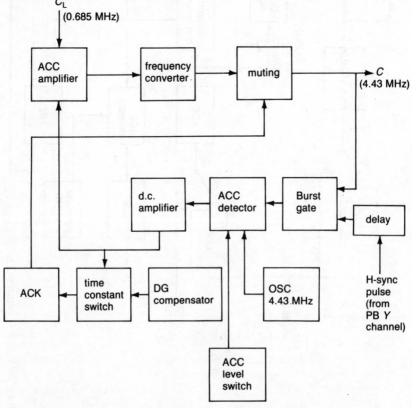

Figure 3.25 The autochromatic control schematic.

operation has been discussed in respect of the REC channel and only the particular points relating to the ACC loop in the PB channel will be explained here.

The ACC level switch (Figure 3.26)

In the special PB modes such as pause, search, slow play and reverse play, etc., the PB level is very weak because the video heads cannot track the corresponding magnetic trace very well. In these modes the low level, i.e. SEARCH/STILL(L) from the control system, will turn transistor Q on and change the ACC detecting level. This means that the sensitivity of the ACC detector is improved in the pause, search, slow and REV modes.

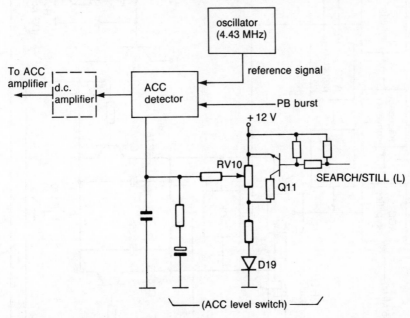

Figure 3.26 The ACC level switch.

The time constant switch (or field storage) circuit (Figure 3.27)

The d.c. error voltages from the ACC detector, without field storage, can only compensate for fast signal variations occurring from line to line. To compensate for long time variations, such as the difference in PB level from field to field caused by the difference in video heads A and B, a field memory circuit for switching time constants has to be used.

Basically, the field memory circuit consists of capacitors C75, C77 and an electronic switch IC7. The RF-SW pulse, in which the high and low levels are assumed to correspond to the operation of video heads A and B,

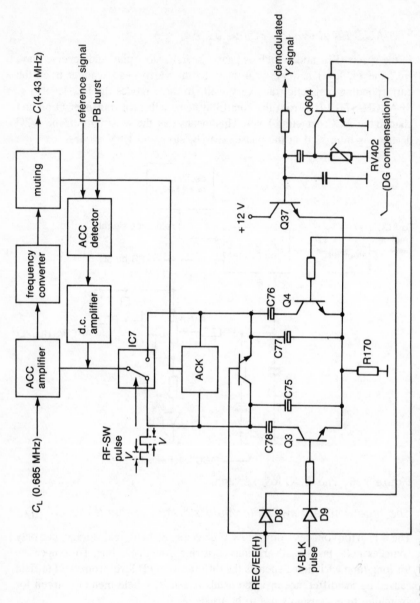

Figure 3.27 The field storage switch circuit.

respectively, controls both capacitors to switch from field to field. A charge stored on each capacitor is proportional to the d.c. average ACC voltage and further affects the gain of the ACC amplifier. Since the burst is blanked for nine television lines in the field blanking interval, the transistors Q3 and Q4 are turned on by the V-BLANK pulse from the servo system. Capacitors C78 and C76, in parallel with C3 and C4, are used to increase the time constant during this period.

Differential gain distortion causes the amplitude of the chroma signal to be amplified more for a high level of Y signal than that of a low level of signal. To compensate for this, the demodulated Y signal is delivered to the base of the buffer Q37, so the voltage across resistor, R107, and the gain of the ACC amplifier will vary with luminance level. A high luminance level will reduce the gain and vice versa. In this way the DG distortion can be reduced to less than 5 per cent. The potentiometer RV402 is used to control the voltage across R and is known as the DG compensation control.

Since problems such as different PB levels from video heads A and B, and DG distortion do not exist in the REC/EE mode, the REC/EE(H) signal is applied to turn on transistors Q3, Q4, Q604 and Q25 thus removing the field memory function and by-passing luminance signal control via Q1. When the chroma signal is very weak or non-existent in the PB signal, the ACK circuit acts and mutes the PB chroma channels.

3.4.3 Time-base error (TBE) and its correction

TBE is a special kind of distortion in VTRs and VCRs because the electromagnetic conversion process is inserted into the video channel between the REC and PB channels. The electromagnetic conversion is performed by head-to-tape scanning or mechanical linkages so any errors in these areas will translate into TBE.

One horizontal period of the standard television signal, e.g. 64 μs for the PAL system, can be used as a reference for measuring variations in the played-back horizontal period. Any instability in this time base can be quantified by measuring TBE. That is to say, the time-base error is the phase deviation, as a function of time, between a standard video signal and a played-back version of the standard video signal, comparing deviations in the horizontal period during playback.

Time-base error in a VCR comprises absolute time-base error (ATBE) and differential time-base error (DTBE), which is the differential of ATBE with respect to time. There is no requirement to minimize ATBE for the PB signal since a lengthened or shortened line period will only result in a stretched or shrunken horizontal image in proportion to line length. Provided that the image size does not change by more than 1 per cent it is unlikely to be perceived by the majority of viewers.

Absolute time-base error will also produce parasitic frequency deviations in the PB FM signal, and a further change in the level of the synchronizing signal and in amplitude deviations in the video signal after demodulation. These effects can be compensated for by clamping and adjusting the output signal.

In order to maintain the recording phase the field synchronizing signal has to be recorded on a specific segment of the track during the recording process. This indicates that ATBE between the time base of the video signal and the position of the video trace on the tape is limited. However, the field synchronizing pulse is quite wide (2.5H = 160 μs for PAL) so strict limiting of ATBE is not necessary and values of $\pm(50-100)$ μs are permitted.

Differential time-base error will make the vertical line on the screen flutter or skew, smear the image edge and reduce the horizontal resolution. It will also produce a random parasitic frequency deviation on the PB FM signal and a further flutter in the amplitude in addition to instability in the contrast of the PB video signal after demodulation. It will also cause phase deviation in the subcarrier resulting in tonal distortion (or saturated distortion for PAL) during chroma signal playback. This means that if DTBE is neglected the picture quality will be degraded.

The head-to-tape scanning system includes the tape transporting system and head drum assembly, and TBE results from these two mechanisms and their servos. The causes of TBE from the tape-transport system are:

1. The manufacturing precision of machining the transport mechanism (the capstan shaft, pinch roller and capstan motor bearing, etc.) cannot be high enough.
2. Parasitic vibration of the transporting tape or belt due to tension or friction changes.

The causes of TBE to changes in rotational speed of the drum are:

1. Vibration of the rotating drum shaft.
2. Deviation of the centre of mass of the rotating shaft from its geometric centre.

Additionally, some deformations of the tape resulting from differences in temperature and humidity between the time of recording and playback, and some disturbances in the drum and capstan servo circuit are also causes of TBE.

Since the linear speed of the video heads on the rotating drum is about 100 times greater than the transport speed of the tape it is usually considered to be the main cause of TBE. To reduce TBE in the VCR, time-base correction is applied in addition to improving the above factors which affect TBE. Time-base error in the PB signal can be completely eliminated using a digital time-base corrector (DTBC) in the VCR, but this equipment is usually very expensive. Time-base correction (TBC) is usually applied to the PB *C* signal to correct the chroma signal only and is present in most VCRs nowadays.

3.4.4 Frequency conversion with TBC

Figure 3.28 shows a general frequency conversion scheme with TBC in the PB channel. Time-base error in the PB chroma signal comprises both frequency component error ($\pm \Delta$) and phase component error ($\pm \delta$) so auto frequency control (AFC) and auto phase control (APC) are included in TBC.

The AFC samples the H-sync signal in the demodulated Y signal and transforms it into an error voltage, $V_{d.c.}(\pm \Delta)$. The APC samples the burst in the reconverted C signal and transforms it into an error voltage, $V_{d.c.}(\pm \delta)$. Both $V_{d.c.}(\pm \Delta)$ and $V_{d.c.}(\pm \delta)$ are applied to a voltage controlled oscillator (VCO) with a centre frequency of 685 kHz (0.685 MHz) for machines with the U-matic format. The output frequency and phase of the VCO are therefore dependent on $V_{d.c.}(\pm \Delta)$ and $V_{d.c.}(\pm \delta)$ resulting in an output related to 685 kHz (0.685 MHz) $\pm \Delta \pm \delta$. In the subconverter this output is added to a reference oscillator signal of 4.43 MHz derived from either an in-built crystal oscillator at 4.43 MHz, or externally from the SC_{IN} socket on the back of the machine. This external source is only used when the machine is connected to a DTBC.

Finally, the C_L signal with TBE $\pm \Delta \pm \delta$, and the output from the subconverter are mixed together to form the reconverted C signal, 4.43 MHz, via a bandpass filter (BPF).

The types of AFC and APC adopted are different according to the model

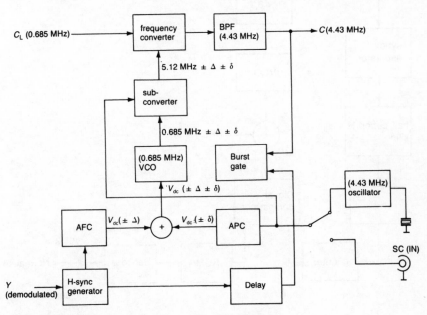

Figure 3.28 Frequency conversion with TBC.

as the AFC. This means that the frequency component error added to the HF of the demodulated Y signal is regarded as a frequency deviation of the FM signal, in which the carrier frequency is the horizontal frequency and becomes $V_{\text{d.c.}}(\pm \Delta)$ after AFC. The phase discriminator (APC) compares the gated burst against the error of the phase component and the reference oscillator signal, mentioned earlier, for phase differences and results in a $V_{\text{d.c.}}(\pm \delta)$ output which is proportional to the phase error between the two.

In the VO-5850P, as shown in Figure 3.29, the principle of the APC is the same as that used in the VO-2860P but the AFC which has been adopted is different. The demodulated Y signal, which comprises the error component of $\pm \delta$, is fed to a monostable multivibrator with $\tau > 32$ μs for killing the

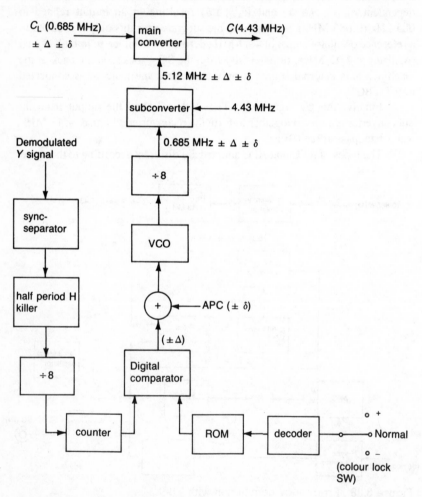

Figure 3.29 APC in the VO-5850P.

pulses for half the horizontal period, T, and then is applied to a counter through a divide-by-eight circuit. Four bits of the binary output from the n counter comprise the error component and are compared with a reference value which is stored in a read only memory (ROM). The address depends on a colour lock switch located on the back of the machine. This switch selects one of three voltages which is translated to the corresponding address of the ROM by a decoder. This data is fed to a digital comparator in which the output voltage depends on the difference between the two. So the output voltage from the comparator comprises the error component of $\pm \Delta$ and is mixed with the output voltage from the APC to be added to the VCO. The centre frequency of the VCO is selected to be eight times 685 kHz (0.685 MHz) for stability improvement and means that the counter counts the horizontal frequency divided by eight. The output from the VCO, which now contains the error components $\pm \Delta$ and $\pm \delta$, is fed to the subconverter via the divide-by-eight circuit and then mixed with a reference oscillator signal of 4.43 MHz and becomes a 5.12 MHz $\pm \Delta \pm \delta$ signal. (The subconverter also consists of a balanced modulator so it is also necessary to include a BPF of 5.12 MHz to filter out the sum frequency signal.) Finally, this 5.12 MHz $\pm \Delta \pm \delta$ and C_L signal (0.685 MHz $\pm \Delta \pm \delta$) are mixed at the main converter and BPF of 4.43 MHz and form the difference frequency of the C_L signal at 4.43 MHz.

3.5 VHS video channel

The video channel includes the Y channel and C channel as shown in Figures 3.30 and 3.31. The record signal path for the Y channel signal, as shown in Figure 3.30, is through the AGC, LPF, clamp, non-linear emphasis, clamp emphasis, white/dark clip, FM modulator, HPF, Y/C mix to the video heads. (Note that the switches are set to the REC or EE side during recording.) The signal path for monitoring sends the signal directly from the AGC to the VIDEO (OUT) socket passing through a PB-EE switch, which is set to the EE side.

The Y signal path during reproduction is from the video heads (R/L) through the R/L heads SW, AGC, DOC, limiter, FM demodulator, LPF, de-emphasis, PB amplifier, LPF, clamp, feedback amplifier, noise canceller and amplifier to the VIDEO(OUT) socket, or the RF converter. These circuits have all been described earlier in relation to the U-matic recorder but apply, in general, to this type of recorder also. (Note that the switches are set to the PB side during reproduction.)

The signal path for recording the C signal, shown in Figure 3.31, is through a BPF to the ACC, frequency converter, REC amplifier, ACK, LPF (< 1 MHz), buffer, producing the C_L signal (to the Y/C mixer). While the signal path for reproducing the C signal is from the video heads (R/L) through the R/L head SW to the LPF, ACC, frequency converter, comb filter (2H),

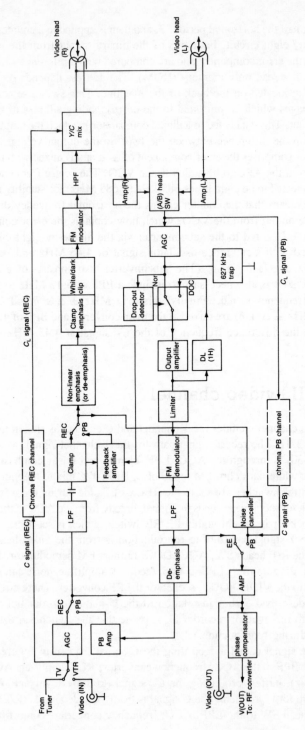

Figure 3.30 The VHS video channel (record path for Y channel signal).

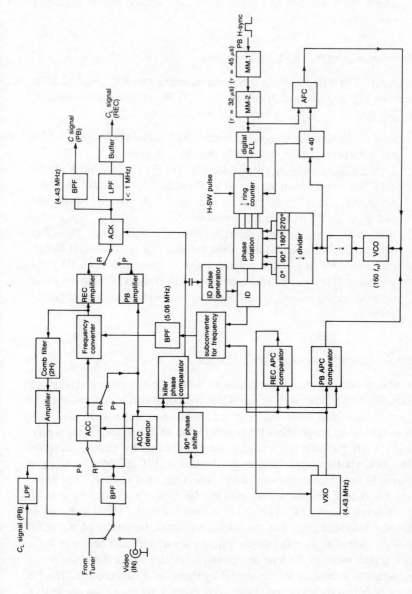

Figure 3.31 The VHS video channel (record path for C signal).

The video channel **87**

amplifier, BPF, PB amplifier, ACK and a BPF (4.43 MHz), after which the C signal is combined with the Y signal and output from the VIDEO (OUT) socket or sent to the RF converter.

There are a number of circuits which are related to the frequency converter.

Auto phase control (APC) circuit

The crystal VCO is free-running during the recording process. The 4.43 MHz continuous signal generated by the crystal VCO is sent to the subconverter. During playback the APC system has three functions:

1. Phase control of the C signal to remove TBE in the PB C signal, which has not been entirely removed by the AFC system.
2. Colour killer control by phase detection. The PB C signal burst and the 4.43 MHz signal from the crystal VCO (with 90° phase shift) are phase compared, and when the phase is normal a high voltage is generated and input to the ACK in the main circuit.
3. AFC control by identical detector (ID) pulse generation. The killer detector is used as an ID and when the phase of the two input signals differ by 180° a positive polarity pulse is generated at an ID pulse generator. Then this pulse is input to the ID in the AFC, and the phase rotation timing errors, which occur at the field switch point, can be corrected.

Auto frequency control (AFC) circuit

The AFC system operates in the same way in both the recording and playback modes. It should, however, be noted that the horizontal synchronization of the input video signal is the AFC standard during recording, and the horizontal synchronization of the playback video signal is the AGC standard during replay. The PB H-sync from the synchronizing separator of the PB Y channel is applied to the AFC via two monostable multivibrators, MM1 and MM2, shown in Figure 3.31, in which the unwanted component has been removed. The phase of the PB H-sync and the signal from the $160f_H$ VCO, divided by 4×40, are compared at the AFC. The output is then fed to the $160f_H$ and used to detect the time-base error and control the oscillation frequency of the VCO. The VCO output is then fed to the $\frac{1}{4}$ frequency divider and becomes $40f_H$, which is then taken to the four-phase rotation circuit where the phase of the $40f_H$ signal is rotated by 90° every 1H by means of a ring counter. The PB H-sync is shifted by a $\frac{1}{4}$ period at the $\frac{1}{4}$ ring counter via the digital phase-lock loop (PLL) and combined with the phase shifted VCO output to change the four $40f_H$ signals by 90° every 1H. Figure 3.32 shows a circuit example which provides phase rotation. The four AND gates in Figure 3.32 are controlled by the output level from the $\frac{1}{4}$ ring counter. A, B, C, D and the $40f_H$ signal with its phase shifted by 0°, 90°, 180° and 270°, which is output from the $\frac{1}{4}$ divider, can be gated by a high level in one of the outputs, A, B, C or D, of the $\frac{1}{4}$ ring counter. Because the period of the output pulses from the ring counter is 4H, i.e. f_H is divided by $\frac{1}{4}$, the output signal from a four input OR gate is $40f_H$ with the phase shifted by 90° every 1H. In addition, because the ring counter is reset by a high level from the head RF-SW pulse, there is a 0° phase signal only (to be output from the four input OR gate) corresponding to the CH-A field. The phase is advanced by 90° in the CH-B field at a high or low level of the H-SW pulse at 25 Hz. The ID inverts the phase of the four-phase rotation circuit output if phase disturbance is caused by the following:

(a)   the switch point for the channel changeover drifts; or
(b)   the rotation of video recording and playback differs.

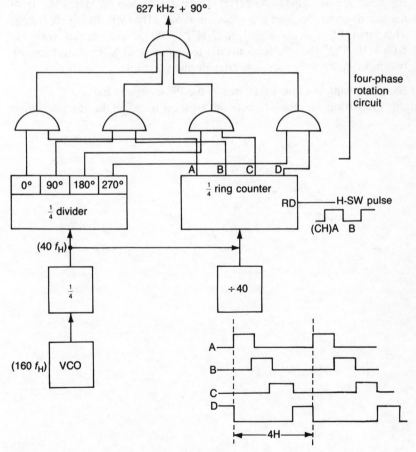

Figure 3.32 PB H-sync phase rotation.

During playback, when playback is started at a different phase from that in recording such as 90° or 270°, the output chroma signal at the point when CH-A changes to CH-B, or CH-B to CH-A, is changed by 180° from the previous phase. When the phase differs by 180° the ID pulse is output from the killer phase comparator in the AFC circuit, inverting the $40f_H$ phase and returning it to the same phase as the original field. When the switching point of the channel selector drifts the same operation is performed.

Frequency converter

The $40f_H$ signal is input into the frequency subconverter with the 4.43 MHz signal from the crystal VCO. The sum frequency component (5.06 MHz) is taken out by the 5.06 MHz BPF and is input into the frequency converter in the main channel. As outlined previously, in the main frequency converter the input chroma signal (4.43 MHz) is mixed with the 5.06 MHz+90° from the subconverter, forming a component at 627 kHz+90° during recording. The input PB chroma signal (627 kHz+90°) is also mixed with the 5.06 MHz+90° from the subconverter to become 4.43 MHz. Thus there are two tasks in the frequency converter during playback:

(a)   removing the time-base jitter in the PB *C* signal; and
(b)   removing the crosstalk between adjacent tracks in the chroma signal.

Chapter 4

The servo system

The servo system, as mentioned in Chapter 2, consists of the drum and capstan servo systems. Each of these are further subdivided into a phase (main) loop and speed (auxiliary) loop. This section discusses the drum and capstan servos of a VCR of the U-matic format using the VO-5850P as an example, and is followed by an example of the servo system in a VHS VCR. The reel servo is relatively simple in a VCR without an edit function, so an example of a reel servo in a VCR with an edit will be described.

4.1 The drum servo system

The rotational speed and phase of the drum motor are controlled by the speed and phase loops in the drum servo system. Some auxiliary circuits are, however, necessary to ensure full operation of the drum servo. These are:

1. The rotating speed of the drum should be changed in the search mode so an AFC branch is necessary in the phase loop.
2. The error in the linear velocity of the video head due to unavoidable drum construction tolerances, such as the use of a brushless commutator motor, drum shaft eccentricity, non-uniform mounting of the eight SPPG magnets, etc., requires compensation by the attachment of a picture splitting compensator to the speed loop.
3. High starting load protection circuits are added to the speed loop of the drum motor.
4. Most of the switches or timing circuits on the channel, framing, system control and servo boards are operated according to the rotating phase of the drum, so some timing pulses are generated from the drum servo system.
5. The relationship between the REF synchronizer and PB synchronizer must be locked during edit before the tape passes through the edit point so a phi-square (ϕ^2) servo is arranged in the phase loop.

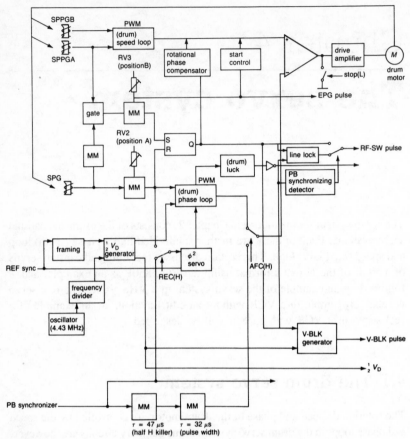

Figure 4.1 Block diagram of the drum servo system.

A block diagram of the drum servo system in the VO-5850P is shown in Figure 4.1.

4.1.1 The phase and speed loops

The principle of the digital servo used in the phase and speed loops has been discussed earlier. A comparison is made between the SPPGA and SPPGB pulses in the speed loop every 5 ms to ensure that the drum motor does one revolution every 40 ms and is rotating uniformly. The SPG pulse is locked with the $\frac{1}{2}V_D$ signal every 40 ms in the phase loop to ensure that the field synchronizing signal is recorded in the appropriate place on the video track.

The output from the phase and speed loops, i.e. the error voltages, are fed, respectively, to the non-inverting and inverting terminals of the operational amplifier, IC10, as shown in Figure 4.2.

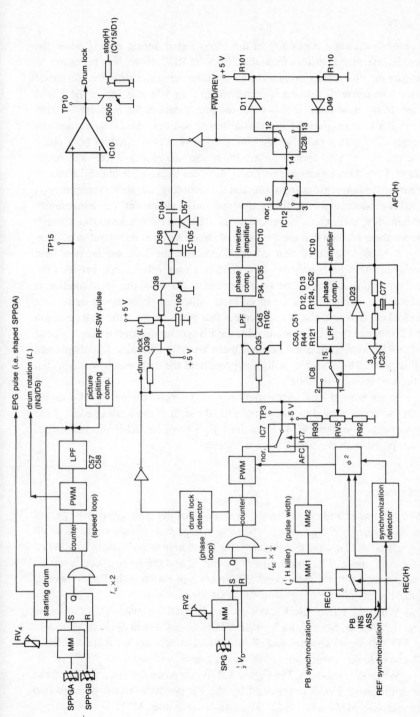

Figure 4.2 The phase and speed loops.

ϕ^2 servo

In order to maintain continuity of the video signal during editing while the recorder machine transfers from the PB to the REC mode, it is necessary to compare the PB synchronizing and REF synchronizing signals in the phi-square servo. The phase difference is converted to data which is fed to the PWM in the drum phase loop. In this case, the preset data in the memory location of the PWM is produced from two different sources, the ϕ^2 servo and the counter. The output pulse width from the PWM is controlled by both phase differences, i.e. SPG and $\frac{1}{2}V_D$, and PB synchronization and REF synchronization. Both these phases are generally different because the phase difference between the PB synchronizing signal and the recording, i.e. REF synchronizing, signal is random, and in addition, the time and environment, i.e. temperature and humidity, is different for recording the REF and PB synchronizing signals. It is for these reasons that the inclusion of the ϕ^2 servo is important in editing.

It should be noted that ϕ^2 lock indicates the lock of the two field synchronizing signals only because there is a time relationship between the field synchronization and line synchronization. So while the drum rotation is uniform, i.e. controlled by the speed loop, both line synchonizing signals are also locked. It should also be pointed out that when the machine changes from the PB mode to the REC mode in edit, the PB synchronization is also replaced by the REF synchronization because there are only EE video signals present at this point. The ϕ^2 servo will, however, hold the data which was detected at the end of the PB mode.

In the normal REC mode there is no PB synchronizing signal present, so the ϕ^2 servo is used to compensate for the shift of the leading edge of the output pulse from the PWM. Here the $\frac{1}{2}V_D$ signal, i.e. REF synchronizing, is compared with the SPG in the ϕ^2 servo.

AFC mode

During the search mode the range of changes of the tape speed is large and the line frequency change is correspondingly large making the monitor difficult to synchronize. To solve this problem, the phase loop is automatically switched to the AFC mode, i.e. auto frequency control, and the drum speed is varied in accordance with the tape speed. The result is to maintain the line frequency of the PB signal into the pull-in range of the monitor.

When the mode select is not set to TBC and the machine is not in the normal PB mode, SLOW, REV or SEARCH modes, but in the PAUSE mode, the AFC(H) signal appears and all three switches shown in Figure 4.2, IC7, IC8 and IC12, are turned on, i.e. the phase loop is switched to the AFC branch.

When the switch, IC7 in Figure 4.2, is turned on the error voltage from the drum phase PWM is replaced by the PB synchronization through two monostables, MM1 and MM2. The first monostable, MM1($\tau = 47 \ \mu$s), is used to eliminate components of $\frac{1}{2}$H, such as the equalizing and the field synchronizing pulses, from the PB composite synchronization, while the second multivibrator, MM2 ($\tau = 32 \ \mu$s), is used to determine the pulse width. This ensures that the amplitude and pulse width of the pulse derived from the PB synchronization are constant and the period only varies with the tape speed in the special tape speed modes.

When the switch IC12 in Figure 4.2 is turned on the lower branch of circuits which comprises the LPF and the phase compensator, etc., are substituted for the upper branch of circuits. The differences between the upper and lower branches are as follows:

1. An LPF with double the time constant is included in the lower branch. This means that when the pulse frequency is higher, i.e. the tape speed is faster, the low time constant can be used for a faster response so the drum speed matches the change of tape speed more efficiently. When the pulse frequency is lower, i.e. the tape speed is slow, the higher time constant can be used to ensure that the drum is more stable at the lower speed.

2. A phase compensator with a large time constant, R124 and C52, is also included in the lower branch to suppress variations in the output d.c. voltage. During the special tape speed modes, the video should be scanning across the guard track resulting in loss of video signal or synchronization signal. If loss of synchronization signals occurs d.c. variations will be increased resulting in the need for the larger time constant.

The voltage divider, which consists of resistors R101 and R110 and the power supply (+5 V), is used to limit the rotating speed of the drum. During the forward search mode the output d.c. voltage from the AFC branch increases with increasing tape speed and results in a drum speed increase, but the drum speed increase is limited by the divided voltage because the $\overline{\text{FWD/REV}}$ instruction from the FWD/REV detector in the capstan servo is (L). Thus the output voltage is clamped when it increases beyond the level of the divided voltage. Conversely, during the reverse search mode the $\overline{\text{FWD/REV}}$ instruction is (H) and the divided voltage clamps the decreasing output voltage through diode D49, so the drum speed in the REV mode is also limited.

To prevent transients in the output voltage when switching the phase loop from the normal PB mode to the search mode, the voltage divider, consisting of R93, R92, RV5 and +5 V, and the delay circuit consisting of D23, C77 and IC23, etc., are arranged to work coincidentally with switch IC8. The divider voltage is adjusted using RV5 to equal the voltage in the AFC branch when working normally. When the machine changes from the normal PB mode to the search mode, the switches IC7 and IC12 are first turned on by the AFC(H) signal, but the switch IC8 is still maintained in the positon connected to the voltage divider because the AFC(H) is delayed by the delay circuit. During this short period the output voltage, which is fed to the drum motor, depends on the present divided voltage from R101 and R110. After this delay the switch IC8 turns on and the output voltage is supplied by the PB synchronizer. (Note that the delay occurs only during switching from the PB mode to the search mode and no delay occurs during reverse switching due to the diode D23.)

Phase loop

In the REV mode the drum phase loop is not locked and during the switch from the REV mode to the FWD × 1 mode, i.e. the normal PB mode, the phase loop needs a short time to transfer from the non-lock to lock position and needs to mute during this period. To do this the $\overline{\text{FWD}}$/REV instruction, mentioned earlier, is also used to control transistor Q35, through IC12, Q38 and Q39. When the transfer from REV mode to FWD mode occurs, a rising edge pulse is generated from the $\overline{\text{FWD}}$/REV instruction through an inverter IC12, and is passed through C104, C105 and D57, D58 (i.e. a double voltage rectifier) and generates a high level. The transistor Q38 turns on, by-passing the charge on C106, thus turning on Q39. The result is that Q35 is supplied from +5 V through Q39 and turns on. The normal branch of the phase loop is muted. After the FWD mode has been set up the constant high level is isolated by the capacitance C104, and Q38 and Q39 are turned off thus removing the muting.

Digital servo

When the number of counting pulses is in the operating range of the counter in the phase loop, the digital servo is working normally and a high level is sent from the drum lock detector. It becomes a low level after passing through an inverter and is referred to as the drum lock (L) signal. The drum lock (L) signal turns Q39 off to remove the muting and also mixes with the capstan lock (L) signal to generate a servo lock signal.

4.1.2 The pulse generator

There are three sorts of pulses generated in the drum servo system. They are:

RF-SW pulse   used to switch the video signal from heads A and B, etc.

V-BLK pulse   used as an artificial V_D signal in the search mode and for indicating the period of the field blank, etc.

E-PG pulse   used in the timing switch for video heads A and B on the BS-3 board.

The block diagram of these pulses is shown in Figure 4.3, and a discussion of these pulses follows.

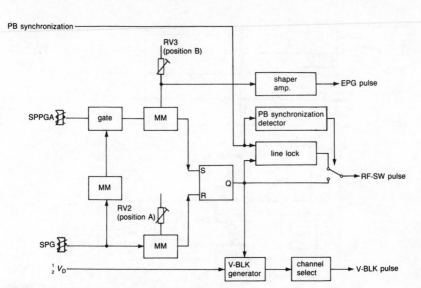

Figure 4.3   The drum servo pulse generators.

RF-SW pulse

The principle of generating the RF-SW pulse is shown in Figure 4.4. The SPG pulse is fed to two monostables to provide a delay. One of the monostables produces a delay of 0.533 ms or 4.8° and is then fed to the R terminal of the R−S trigger so that the R−S trigger is reset at switch point A, while the other one produces a delay of 19.14 ms, or phase shift of 172.3°, and is used to gate one of the SPPGA pulses. The last SPPGA pulse before switch point B is gated and is delayed, by a following monostable, to that switch point. Using this pulse, located at switch point B, to set the R−S trigger, a symmetrical square wave, of period 40 ms and pulse width 20 ms, is output at the Q terminal of the R−S trigger. The two edges of this square wave appear at the switch points A and B, respectively. This pulse is known as the RF-SW′ pulse.

In the REC mode, the SPG pulse and the $\frac{1}{2} V_D$ signal are locked, with the $\frac{1}{2} V_D$ signal originating from the REF synchronization, i.e. the RF-SW′ pulse. In this case it can be sent out directly and is referred to as the RF-SW pulse. However, in the PB mode the $\frac{1}{2} V_D$ signal is generated from the internal oscillator so, after it is generated from the SPG pulse, the RF-SW′ pulse is locked in the television line by the PB synchronizer as shown in Figure 4.5. In this manner the edge of the RF-SW′ pulse can be shifted to exactly 6.5H before the leading edge of the field synchronization. The switch shown in Figure 4.4 is turned on only when the PB synchronization appears in the PB mode. The generated RF-SW pulse is also sent to the following circuits:

(a)   to reset the timing generator of the switch for the video heads A and B;

(b)   to be used for switching the PB video signals from video heads A and

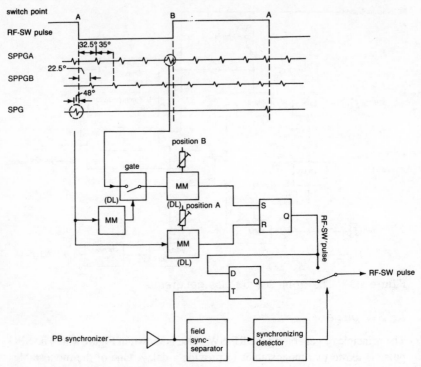

Figure 4.4 RF-SW pulse generator.

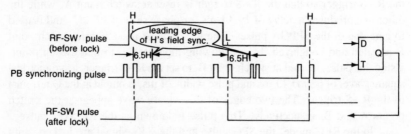

Figure 4.5 Locking of the RF-SW′ pulse.

B, and for detecting the drop-out pulse at the DOC detector; and

(c) on the rotational phase compensator and the V-BLK pulse generator in the drum servo to be used to generate the compensating voltage and the V-BLK pulses, respectively.

V-BLK pulse

For discussion purposes the V-BLK pulse can be considered to be similar to the RF-SW pulse, i.e. the V-BLK′ pulse is first generated then the V-BLK pulse generated. Figure 4.6 shows the generation of the V-BLK pulse. The RF-SW pulse and the $\frac{1}{2} V_D$ signal are fed to an exclusive OR gate and the output is a pulse in which the period is 20 ms and the pulse width is an interval between the edges of the two input pulses. This pulse is the V-BLK' pulse. It can be seen from Figure 4.6 that the V-BLK' pulse passes through three paths to TP31. This is necessary to allow for the various replay modes. These are:

Path 1 QA to R505 to Q501 to Q6 to TP31
Path 2 QA to R506 to D502 to IC25-4 to Q6 to TP31
Path 3 QA to IC25-12 to IC25-4 to Q6 to TP31

It should be pointed out that, in the REC mode the output from Q4, i.e. the V-BLK' pulse, is by-passed by the EE (L) signal, from the control system, so no output is present at TP31.

For the convenience of further discussion, it is worth reviewing the following instructions from the microcomputer control system.

STILL/AFC(H)	This is high when in the pause mode
SEARCH(H)	This is high when in the search mode, including the slow mode and the REV mode
PINCH ON (L)	This is low when the pinch solenoid is turned on
TBC(L)	This is low when the 'mode select' on the front panel is set to TBC

When in normal PB mode, i.e. PINCH ON is low and $\overline{\text{SEARCH}}$ is low, Q502 is turned off and the V-BLK' pulse is passed through path 1 to TP31 and is used as the V-BLK pulse at this point.

When in the pause mode, i.e. STILL/AFC is high, Q503 turns on, and Q501 and Q502 turn off, path 1 is inhibited and path 3 is selected. In this case IC25-12 of the monostable is triggered by the trailing edge of the V-BLK' pulse, i.e. the edge of the $\frac{1}{2} V_D$ signal. After a delay of 0.2 ms at the first monostable, the pulse width of 5H is produced at the second monostable and is referred to as an artificial field blank pulse. This occurs because, in the pause mode, the period of the RF-SW pulse changes with the drum rotational speed, but the $\frac{1}{2} V_D$ signal period does not change. The result is that the V-BLK' pulse is also changed, so it has to be replaced by an artificial field blank pulse which can be output from TP31.

When the machine is connected to the time-base corrector and the 'mode select' is set to TBC, the artificial blank is not necessary and path 3 is inhibited by TBC(L).

When in search mode (i.e. SEARCH is high), Q502 turns on and not only turns Q501 off (i.e. inhibits path 1), but also resets the first monostable (i.e. inhibits path 3). However, because IC25-9 is high at this time, D502 is turned on, i.e. the V-BLK' pulse can pass along path 2. In other words, generation of the artificial field blank pulse is also necessary in the search mode.

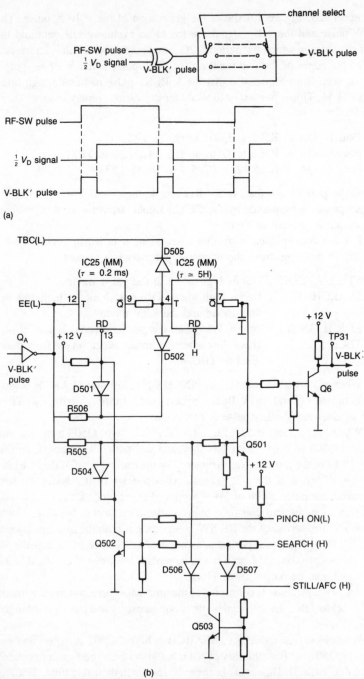

Figure 4.6 Generation of the V-BLK pulse: (a) V-BLK′ pulse generator, (b) V-BLK pulse generator.

The generated V-BLK pulse is sent to the PB *C* channel, ACC and APC for indicating a blank period. It is also used as an artificial V-BLK pulse at the output circuit of the PB channel during the search and pause modes.

E-PG pulse

The E-PG pulse is very simple to generate: it is output from the SPPGA pulse through a pulse shaping amplifier. Then it is sent to the control system for use in the timing switch for video heads A and B.

4.1.3 Rotational phase compensation and control

The rotational phase compensator

This is also referred to as the picture splitting compensation and is used to compensate for the error caused by non-uniform rotation in each revolution. These errors are cumulative, i.e. the error from the second revolution is based on the first one, so it should become greater during recording or replay. This error will result in TBE and must be compensated for, especially in an editing machine. An anti-phase correction method has been adopted for this compensation, as shown in Figure 4.7a.

After an inverter, Q16, the rising edge of the RF-SW pulse triggers the first monostable. The output pulse width at the $\overline{Q}$ terminal of this monostable can be adjusted by RV6. The second monostable is triggered by the rising edge of the first output pulse. Both pulses output from Q and $\overline{Q}$ of the second monostable are equal in amplitude and of opposite polarity. The amplitude of the second output pulse can be adjusted by RV7. This output pulse will become a compensating voltage via an integrating circuit.

We can summarize by saying that the adjustment of RV6 determines the place which needs to be compensated and the adjustment of RV7 determines the compensating amplitude. Since the polarity of the output pulse from RV7 can be either positive or negative, dependent on whether RV7 is adjusted up or down, the position of the compensation can be over the range 0° to 360°.

Control circuits

Control circuits arranged in the speed loop of the drum servo include the drum start circuit and the drum protection circuit as shown in Figure 4.8. Before the drum motor starts there is no SPPGA pulse to the PG pulse amplifier, so transistor Q504 is turned off as its base is connected to ground. At the same time the +12 V supply is connected to IC10-3, via diode D510, as a starting voltage for the drum motor.

After the drum starts and rotates normally, pulses SPPGA and SPPGB

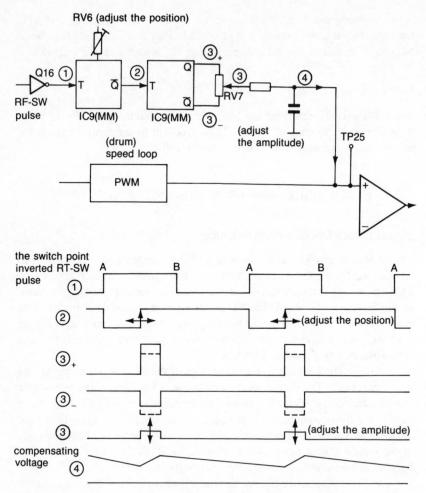

Figure 4.7 Rotational phase compensation and control.

are generated and the SPPG pulse passes through the PG amplifier to a voltage doubler comprising D508, D509, C503 and C504. The high voltage level produced turns Q504 on and, thus, D510 off. The speed loop is then working normally. A low logic level is sent from the PWM in the speed loop to the microcomputer control system and is used to execute a programme for the normal operation of the drum motor.

When some problems occur to prevent the starting of the drum motor, the above input control signal from the PWM to the microcomputer changes from low to high logic level. This is an auto stop instruction, which makes the computer execute a stop programme. This output instruction is changed to a high level through an inverter and turns on transistor Q505, as shown in Figure 4.8. The result is to by-pass the drive voltage of the drum motor causing the motor to stop.

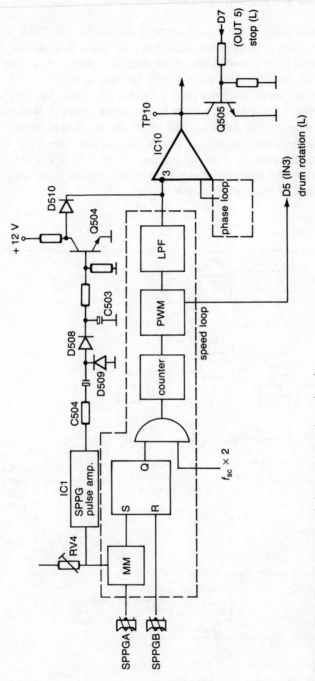

Figure 4.8 Drum start and drum protection circuits.

The servo system **103**

4.2 The capstan servo system

The block diagram of the capstan servo system in the VO-5850P is shown in Figure 4.9. It consists of the speed loop, the phase loop and the circuits which control the rotational speed, direction and the pause servo. The digital servo circuits in the capstan servo system are used in exactly the same way as in the drum servo described previously. Because speed and direction are mainly controlled in the search mode by the capstan servo, it is necessary to arrange a detector and control circuit which operate in association with the microcomputer control system, to control the rotational speed and direction of the capstan motor. In addition, the pause servo is designed to locate the guard band at the top or bottom area of the screen picture in the pause mode, or in the still position of the search mode.

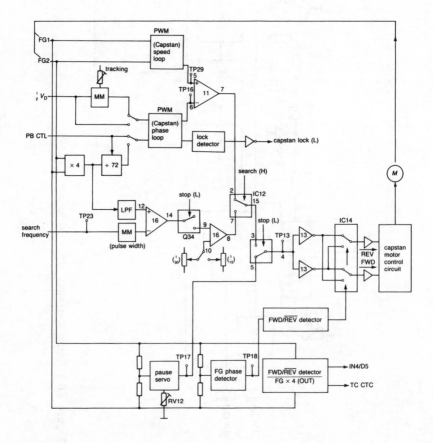

Figure 4.9   Capstan servo system.

4.2.1 The phase and speed loops

Pulses FG1 and FG2 are compared 450 times per second in the speed loop, while the phase loop locks the $\frac{1}{2}V_D$ signal and the FG signal or the PB CTL signal, according to the selected mode, every 40 ms to ensure either maintenance of a stable tape speed during the REC mode, or accurate tracking of the recorded video trace on the tape during the PB mode. This is shown in Figure 4.10.

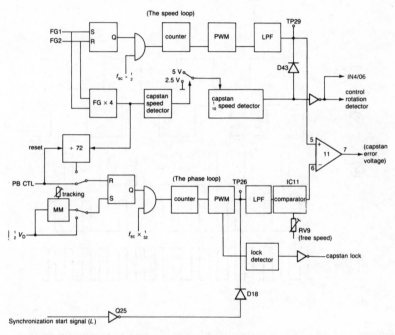

Figure 4.10 The phase loop and speed loop.

The FG1 and FG2 pulses are also fed to a frequency multiplier ($\times$ 4), and to a FWD/REV detector, except in the case of the capstan speed loop. A simple $\times$ 4 frequency multiplier can be produced using two frequency doublers and an XOR gate as shown in Figure 4.11. Because a phase difference exists between the FG1 and FG2 pulses (it is 90° in the FWD mode) the double frequency signals can add up to a $\times$ 4 signal at the XOR gate. The FG $\times$ 4 signal is split up into four paths:

(a) the capstan speed detector;
(b) the capstan phase loop;
(c) the path used in the search mode of the capstan servo; and
(d) the capstan $\frac{1}{2}$ speed detector.

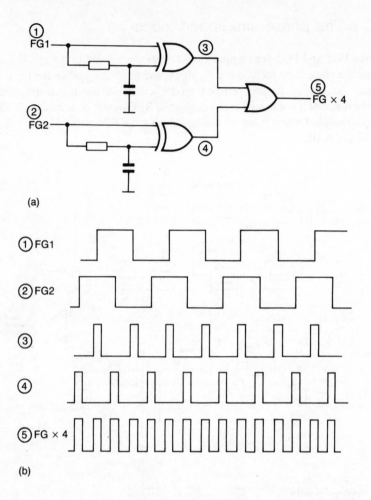

Figure 4.11 A ×4 frequency multiplier: (a) one of the methods for generating an FG × 4 signal, (b) waveforms. Note: the FG1 and FG2 signals have been shaped.

Since no FG signal exists before the capstan starts in the period between STOP and PLAY, a high level (+5 V) is sent from the capstan speed detector to the output of the capstan speed loop (TP29) through D43 until the tape speed is greater than $\frac{1}{10}$ normal tape speed. The signal detected at capstan $\frac{1}{2}$ speed is used as an input signal to the microcomputer when the capstan speed is greater than $\frac{1}{2}$ of normal tape speed. The other two paths will be discussed in detail later.

In the phase loop the input signals are selected according to the different modes as shown in Figure 4.12. We know that when the PB CTL signal appears in the normal PB or INS edit modes, it can be used to lock with the $\frac{1}{2} V_D$

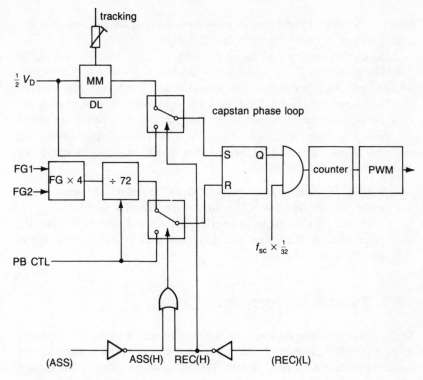

Figure 4.12 Capstan phase loop.

signal; and when the PB CTL signal does not appear in the normal REC mode or ASS edit modes, the FG signal is used to lock with the $\frac{1}{2} V_D$ signal. We also know that the $\frac{1}{2} V_D$ signal needs to be controlled to allow tracking in the PB mode. In both the ASS and INS edit modes, PB is selected before the edit-in point is passed and control of the $\frac{1}{2} V_D$ signal is also necessary.

To summarize, the input signals should be selected as follows for the capstan phase loop:

$\frac{1}{2} V_D$ (via tracking control) $\longleftrightarrow$ PB CTL in the PB or INS mode

$\frac{1}{2} V_D$ (via tracking control) $\longleftrightarrow$ FG in the ASS edit mode

$\frac{1}{2} V_D$ $\longleftrightarrow$ FG in the REC mode

The circuit shown in Figure 4.12 is designed to meet these requirements.

In Figure 4.12 the upper switch is controlled by the REC(H) signal so the $\frac{1}{2} V_D$ signal can be selected in the REC mode. The $\frac{1}{2} V_D$ signal, via tracking control, can be selected in the PB, ASS or INS modes. The lower switch is controlled by both the ASS(H) and REC(H) signals, so that the FG signal can be selected in the REC mode or the ASS mode. The PB CTL signal, however, is selected to the capstan phase loop in either the REC mode or the ASS mode, or in either the PB or the INS mode.

Since the frequency of the $\frac{1}{2}V_D$ signal is 25 Hz the FG × 4 signal should be divided by 72, i.e. $450 \times 4/72 = 25$ Hz, so it is necessary to explain why the frequency of the FG signal should first be multiplied then divided. This is because the signal locking the $\frac{1}{2}V_D$ signal needs to change from the PB CTL signal to the FG signal when changing from playback to record at the edit-in point in the ASS mode. The $\frac{1}{2}V_D$ signal takes a longer time to pull in the FG signal and picture flicker occurs at the edit-in point. There are two methods of avoiding this effect:

1. Raising the frequency of the FG signal to reduce the static phase difference between the FG signal and the PB CTL signal.
2. Using the PB CTL signal to reset the FG signal divider before the edit point is passed. This ensures that the phases of the FG and PB CTL signals are identical at the edit point.

4.2.2 Capstan servo system

The circuits controlling the rotational speed and direction of the capstan motor in the VO-5850P are mainly used in the search mode and are partly served by the microcomputer control system. These circuits are located in the capstan servo circuit and are shown in Figure 4.13.

Capstan speed

The circuit controlling the capstan speed is switched at IC12 and controlled by a SEARCH(H) signal. When the REV mode, the still mode or the search mode is selected, the SEARCH(H) signal is high and a search frequency can be selected to drive the capstan motor by passing through switch IC12. This search frequency is a string of controllable pulses from the control system, originating either from the auto-editor, RM-440, or generated from the machine itself. By controlling a SEARCH dial on the front panel of the machine, or by executing an edit programme automatically, the search frequency can be controlled. For example, turning the SEARCH dial with an angle encoder, four data, D_0, D_1, D_2 and D_3, will be generated from the encoder. The four data correspond to sixteen positions, or angles, of the turned dial (note that $2^4 = 16$). This is equivalent to eight tape-speed selections in both the FWD and REV modes. These correspond to $\pm(\times 5)$, $\pm(\times 2)$, $\pm(\times 1)$, $\pm(\frac{1}{2})$, $\pm(\times \frac{1}{5})$, $\pm(\times \frac{1}{10})$, $\pm(\times \frac{1}{30})$ and $\pm(\times 0)$. Of the four data, D_3 is a FWD/$\overline{\text{REV}}$ instruction to control the capstan direction, while the other three data, D_0, D_1 and D_2, are used to determine the eight time constants in channel 2 of the counter/timer controller (CTC) and then eight search frequencies from

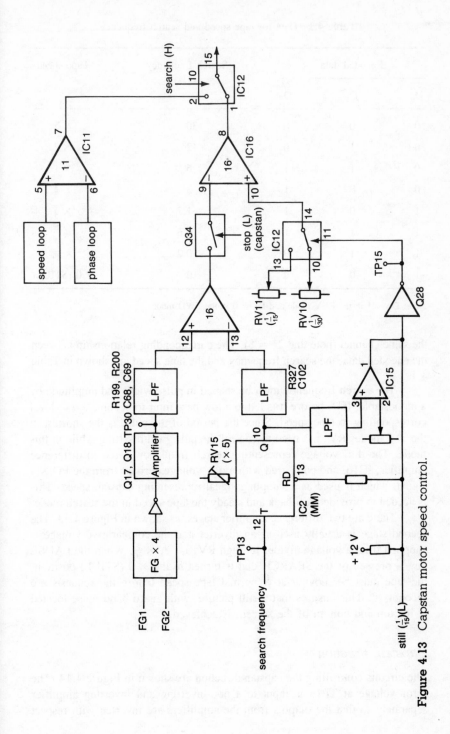

Figure 4.13 Capstan motor speed control.

Table 4.1 Data for tape speed and search frequency

Encoded data			Search frequency (kHz)	Tape speed
D_0	D_1	D_2		
0	0	0	30	$\times 5$
0	1	0	12	$\times 2$
0	1	1	6	$\times 1$
0	0	1	3	$\times \frac{1}{2}$
1	0	1	1.2	$\times \frac{1}{5}$
1	1	1	0.6	$\times \frac{1}{10}$
1	1	0	0.2	$\times \frac{1}{30}$
1	0	0	0	$\times 0$ (STILL)

Note: $D_3 = 1$ in the REV mode and $D_3 = 0$ in the FWD mode.

the same channel (note that $2^3 = 8$). The corresponding relationship between the encoded data, the search frequency and the tape speed are shown in Table 4.1.

The search frequency must be shaped in pulse width and amplitude by a monostable, IC2, before it is fed to a low pass filter to become a d.c. level corresponding to tape speed. Since the period of the pulse is the shortest in the $\times 5$ mode, RV15 is used to control the pulse-width setting while in this mode. The d.c. voltage representing search frequency is fed to difference amplifier, IC16, and compared with a d.c. voltage derived from the FG $\times 4$ signal, which is used as a sampling signal representing capstan speed. This is needed to provide feeedback and steady the tape speed in the search mode.

There are two difference amplifier stages, as shown in Figure 4.13. The second stage is usually used as an inverter and the comparative voltage is supplied from a voltage divider through RV10, i.e. $\times \frac{1}{30}$. When the PAUSE key is pressed or the SEARCH dial is turned to the $\times 0$ (STILL) position, the tape must be moving at $\frac{1}{15}$ normal tape speed before the requests are recognized. This ensures that a still picture, with guard band noise located at the top and bottom of the screen, is achieved.

Capstan direction

The circuits controlling the capstan direction are shown in Figure 4.14. The error voltage at TP13 is input to a non-inverting and inverting amplifier separately so that the outputs from the amplifiers are inverted with respect

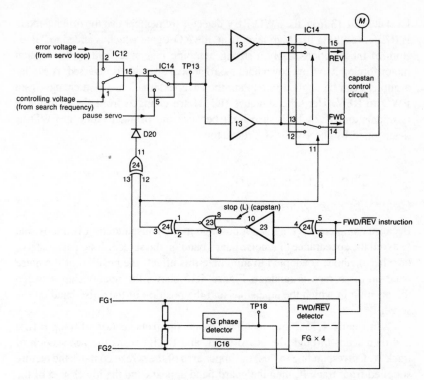

Figure 4.14 Capstan direction control.

to each other. When the level at IC14-11 is high, the direction selection switches (IC14) are connected to the A terminals, i.e. the current flows from IC14-14 to IC14-15 through the capstan control circuit, and the capstan direction is in the FWD mode.

When the level at IC14-11 is low, IC14 is switched to the B terminals so the current flows in the reverse direction and the capstan is rotating in the reverse direction, as in the REV mode. The level at IC14-11 is determined by the control circuit, while the control circuit consists of IC23 and IC24.

The case which occurs when the STOP(L) signal appears will be discussed in the following section dealing with the pause servo. For the current discussion, IC23-8 is assumed to be at a high logic level and IC23-10 is low. If the mode is changed from REV to FWD, the original mode is REV, i.e. the switches IC14 are connected to the B terminals and the level sent from the FWD/$\overline{\text{REV}}$ detector is low.

When the FWD instruction appears IC24-2 is high and then IC24-3 is high. This high level at IC24-3 not only changes the switches (IC14) from the B to the A terminals, but is passed to IC24-12 as well. The high level changes the flow direction and generates a damping magnetic field in the capstan, while the signal at IC24-12 is applied to an XOR gate with a low level at IC24-13 from the FWD/REV detector (remember that the original mode is REV), resulting in a high level from the XOR gate which is added to TP13. Both of these actions apply a strong damping magnetic field to the capstan motor forcing it to slow down until the rotational direction is reversed. A similar argument can be applied to explain the reverse situation, i.e. a change from FWD to REV. Here the switches (IC14) are changed from the A to the B terminals and IC24-11 is also high because the original mode is FWD as determined by the FWD/REV detector.

4.2.3 The PAUSE servo

Control of noise guard band

Because complete tracking is not possible in the pause mode, it is impossible to avoid the appearance of a noise guard band on the screen. If no pause servo, or other method, is adopted to minimize this effect, the position of the noise band on the screen is random. Figure 4.15 shows the relationship between the position in which the tape stops and the position of the noise band on the screen.

In Figure 4.15a, the video head scans the front section of track B first and then scans through the guard band, and finally scans the last section of track A. Corresponding to this, the upper area of the screen contains the picture supplied from track B, then the guard band appears and the lower area of the screen contains the picture from track A.

As shown in Figure 4.15b, the video head scans the guard band first and last, while the middle section scanned by the video head originates from track A. The picture from track A is located in the middle of the screen, while the guard band appears at the top and bottom of the screen. This obviously is the preferred case.

Figure 4.16 shows five positions in a period of the CTL signal in which the tape can be chosen to stop in the pause mode. Positions 1, 3 and 5 are shown in Figure 4.15a, while positions 2 and 4 are shown in Figure 4.15b. The FG signal is a sampling signal of the rotational speed of the capstan so it is fed to the microcomputer system to be used as a counting pulse in the pause mode.

We know that the frequency of the FG signal is 450 Hz, so the FG $\times$ 4 signal frequency is 1800 Hz and the rotational speed of the drum is 25 turns per second, therefore $1800/25 = 72$ pulses are counted each time the drum rotates one turn, or for one period of each CTL signal. Positions 2 or 4 (shown in Figure 4.15b) correspond to $\frac{1}{4}$ or $\frac{3}{4}$ of a period of the CTL signal, so if a pause operation is requested at this point a STOP(L) signal will be generated by the microcomputer when the counter counts $\frac{1}{4} \times 72 = 18$, or $\frac{3}{4} \times 72 = 54$ pulses. This ensures that the counter is reset at the beginning of a CTL period.

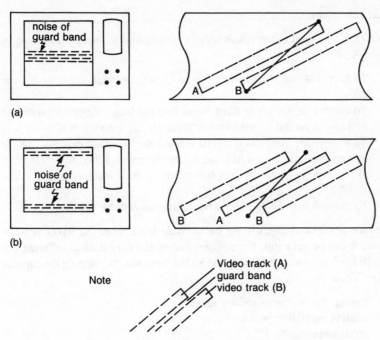

Figure 4.15 The relationship between guard band and tape stop position.

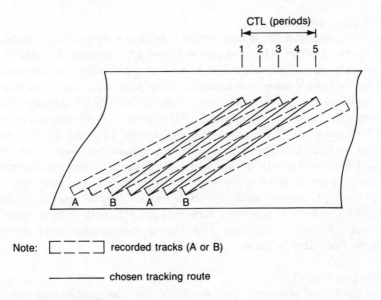

Note: [⬚⬚⬚] recorded tracks (A or B)

——————— chosen tracking route

Figure 4.16 Tape stop positions in one CTL period in the PAUSE mode.

STOP(L) signal

The STOP(L) signal at the pause servo controls four elements, as shown in Figure 4.17.

1. To turn Q34 off, and then the output from the capstan servo loop is cut off.
2. To control the switch at IC14-9 and then the stop voltage, generated at TP17, can be fed to the capstan drive circuit via switch IC14.
3. To remove the inhibit at IC23-10 to allow the FG phase detector (TP18) to combine with the circuit controlling the capstan direction. The stop of the capstan motor is then controlled by both voltages at TP17 and TP18.
4. To control two AND gates whose function will be detailed later.

When the machine changes to the pause mode from either the FWD or REV modes, it can be split into three stages during the period from pressing the PAUSE key, or turning the dial to the STILL position, to stopping the capstan motor. These are:

(a) forcing the motor to reduce speed;
(b) rotating inertially; and
(c) servo stopping.

These steps are discussed below.

REDUCING SPEED

The step forcing the motor to reduce speed is the same as the process controlling the capstan direction. This is shown in Figure 4.14. Because the STOP(L) signal has not appeared at this step, IC23-8 is high and the microcomputer sends out the FWD/$\overline{REV}$ instruction according to the pause input instruction from the keyboard and the detected signal from the FWD/$\overline{REV}$ detector. For example, for the mode change from FWD to pause, i.e. the original mode is FWD, the FWD/$\overline{REV}$ instruction is low and the STOP(L) signal is not present, so IC23-8 is high or IC23-10 is low. Both of these low signals are applied to the XOR gate (IC24) and then IC24-3 is low. This low level controls the direction switch, IC14, to generate a damping magnetic field in the motor. It is also fed to another XOR gate (IC24) with the signal detected by the FWD/$\overline{REV}$ detector to generate a high level at TP13. Both of these actions will control the motor in reducing speed. The explanation applies to the change of mode from REV to pause.

INERTIAL ROTATION

When the capstan rotational speed reduces to less than $\frac{1}{10}$ of normal speed, the transistor Q14, or Q15, will be turned on by the output of the speed detector, so the damping voltage is shorted out and the motor continues rotating under

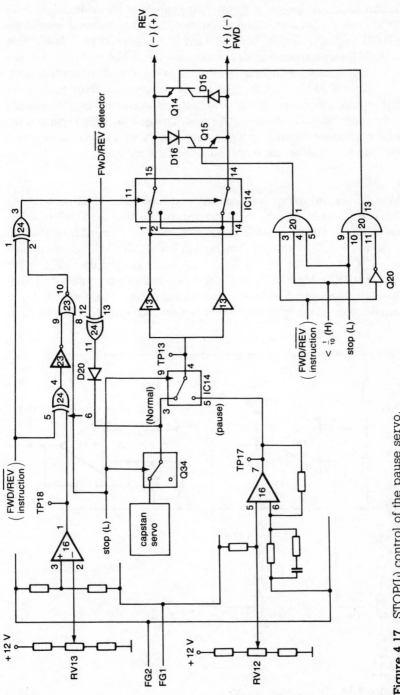

Figure 4.17 STOP(L) control of the pause servo.

The servo system **115**

inertial forces. As before, if we take the example of the mode change from FWD to pause, because IC20-3 is low and IC20-11 is high and because the STOP(L) signal is not present, IC20-5 and IC20-9 are both high. In addition, the level from the speed detector changes from low to high when the rotational speed is less than $\frac{1}{10}$ of normal speed so IC24-4 and IC20-10 are both high.

Next, IC20-13 is high and Q14 is turned on, which just shorts the damping high voltage (i.e. it is positive polarity at the lower side and negative polarity at the upper side). The discussion for mode change from REV to pause is the same and because the damping high voltage polarity is the reverse of that above and Q15 is turned on, the output voltage is also shorted.

STOPPING

After the damping voltage is shorted, the motor continues inertial rotation until the sixteenth input pulse has been counted. By this time the STOP(L) signal has appeared. For the example in which the mode is changed from FWD to pause mode, IC20-9 changes from high to low, IC20-13 changes from high to low, Q14 is turned off and the capstan motor is again controlled by the voltage at TP13. Meanwhile, the STOP(L) signal also cuts off the output from the capstan servo loop, Q34 is turned off and changes the switch, IC14, to stop the motor at the optimum position as mentioned earlier. This takes place

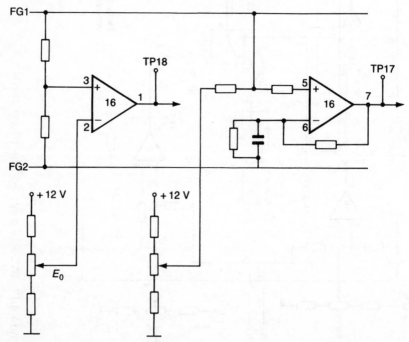

Figure 4.18 Test points 17 and 18.

under the control of the voltages at TP17 and TP18. In fact, IC14-4 is switched by the STOP(L) signal to IC14-5, so the voltage at TP13 is supplied by TP17. In addition, the voltage at TP18 is combined with the FWD/$\overline{\text{REV}}$ instruction to control the direction selection switch, IC14, because IC23-8, the STOP(L) signal, is low.

The circuits which generate the voltages at TP17 and TP18, and their waveforms, are shown in Figures 4.18 and 4.19. Two operational amplifiers

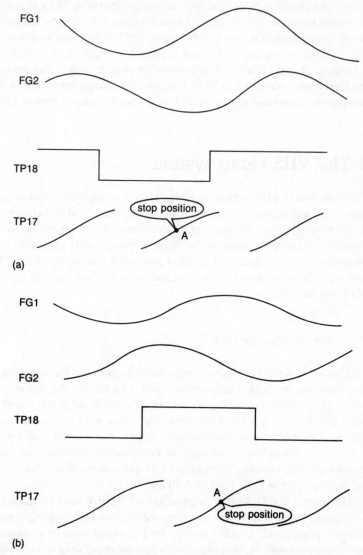

Figure 4.19 Waveforms at TP17 and TP18: (a) in FWD mode (FG1 advances FG2), (b) in REV mode (FG1 lags FG2).

generating the voltage at TP18 are operating at saturation, i.e. the output at TP18 is high when the instantaneous voltage of FG1 + FG2 is greater than a comparison voltage E_0, and the output at TP18 is low in the reverse case. The operational amplifier generating the voltage at TP17 is acting as an amplifier, i.e. the output at TP17 is an amplification of FG1 − FG2. Because the original mode is FWD for the case of FWD to pause, the direction selection switch, IC14, is on the A terminals, i.e. IC14-1 and IC14-13 are connected together, so only when IC24-3 is low, or the waveform at TP18 is low, is the direction selection switch changed over to connect the reverse terminals. A slowly rising voltage is sent to TP13 from TP17 and is used as a braking voltage to make the capstan motor stop at point A in Figure 4.19. Again, the case for mode changes from REV to pause is the same as above. For this case a high level output waveform at TP18 is required to change the switch, IC14, and the capstan motor also stops at point A of the slowly rising voltage at TP17.

4.3 The VHS servo system

The servo system in VHS is the same as that in the U-matic system and consists of both drum and capstan servo. It includes both phase and speed loops in each servo system. Block diagrams of the drum and capstan servos are shown in Figures 4.20 and 4.21, respectively. Because the operating principle of the servo system has been discussed in detail previously, the servo of the VHS system will only be discussed briefly here and only additional contents relating to VHS will be added.

4.3.1 The digital servo

The digital servo is used in all servo loops for both drum and capstan control. During recording the field synchronizing signal, which originates in the video channel and is referred to as V_{SS}, is split three ways in the servo circuit in Figure 4.21. One path is to the drum phase loop, where it is used as a reference signal (the $\frac{1}{2} V_D$ signal mentioned earlier) where it is compared with a pulse from FG1, i.e. a sampling signal from the drum motor. The other two paths are to the capstan servo after generating a CTL pulse in the drum servo. This signal has the symbol C in Figures 4.20 and 4.21.

The generated CTL signal is applied to the CTL R/P head for recording and compared with a sampling signal from the capstan FG sampling head in the capstan phase loop. In other words, the V_{SS} signal used as a sampling signal is compared with the sampling signals from the drum and capstan motors in the respective phase loops. The V_{SS} signal is converted into a CTL signal and recorded on the tape.

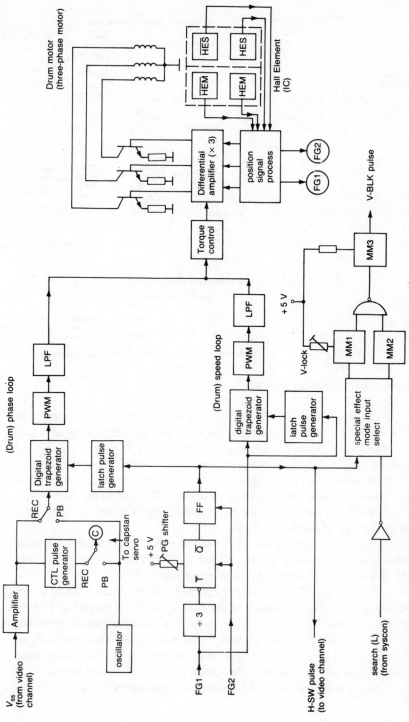

Figure 4.20 VHS drum servo system.

The servo system **119**

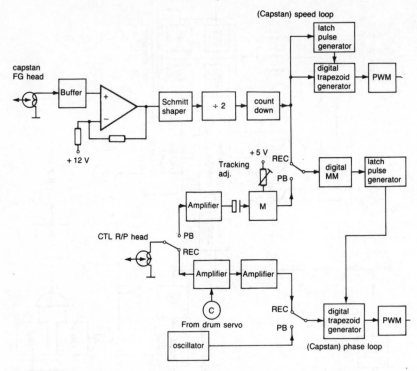

Figure 4.21 VHS capstan servo system.

During playback the PB CTL signal is first compared with a signal from a local oscillator in the capstan phase loop, and then the oscillator signal, which has been locked in phase and frequency with the PB CTL signal, is compared with the sampling signal from the drum motor in the drum phase loop. In both the drum and capstan speed loops the method is used in which a sampling pulse, i.e. the FG signal in the capstan servo or the FG1 signal in the drum servo, is compared with itself. The principle of operation in these schemes is the same as that described in Chapter 2.

The circuits which generate the H-SW pulse and the V-BLK pulse are also the same as those described for the U-matic VCR. Figure 4.22 shows the H-SW pulse generator and its waveforms.

The FG1 and FG2 signals are generated from the drum motor as shown in Figure 4.21. The frequency of the FG2 signal is 25 Hz and is $\frac{1}{6}$ of the frequency of the FG1 signal. The divide-by-three circuit divides the frequency of the FG1 signal following its reset by the FG2 signal. The divided signal is then delayed by a monostable to ensure that the leading and trailing edges of the signal are lined up with the switch points of the video heads.

Figure 4.23 shows the V-BLK pulse generator and its waveforms. During

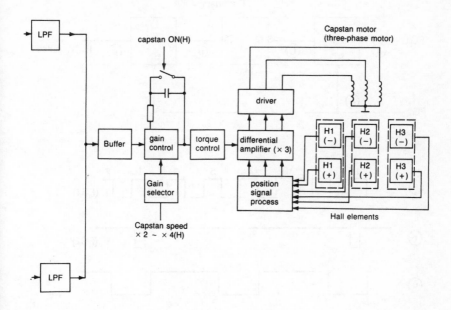

the search mode, a low level from the control system is inverted by an inverter and applied to one input of a NAND gate. The H-SW pulse at the other input of the NAND gate is then permitted to pass through, although inverted. This inverted pulse then goes in two directions. One is through an inverter to a monostable, MM2, and the other is applied directly to a monostable, MM1. Both monostables are triggered by the rising edge of the input pulse resulting in a negative polarity output pulse as shown in waveforms (c) and (d) in Figure 4.23. Both pulses are inverted at a NAND gate and are applied to a monostable, MM3, to form a pulse width at the field period of 20 ms.

4.3.2 Drum and capstan motors

The drum and capstan motors, shown in Figures 4.20 and 4.21, are different from those in the U-matic VCR described earlier. VHS uses three-phase motors in which the current in the three coils of the motor is generated by three different signals, $V_1(\phi)$, $V_2(\phi)$, $V_3(\phi)$. The three waveforms are shown in Figure 4.24, where $\phi = \omega t$ and ω is the angular frequency of each of the three signals.

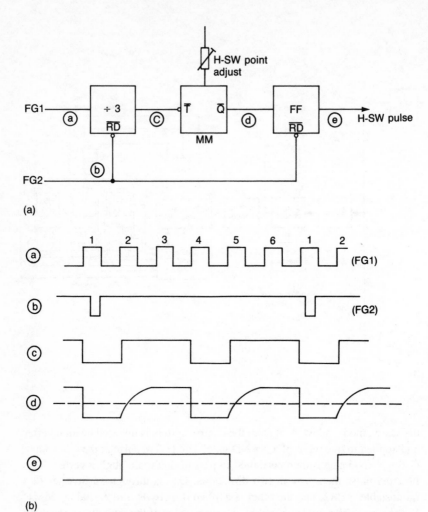

Figure 4.22 The H-SW pulse generator and its waveforms: (a) H-SW pulse generator, (b) waveforms of (a).

In Figure 4.24a:

$$i_a = \frac{V_1 - V_2}{2Z}$$

$$i_b = \frac{V_3 - V_1}{2Z}$$

$$i_c = \frac{V_2 - V_3}{2Z}$$

where Z is the impedance of each coil and the impedances of all the coils are

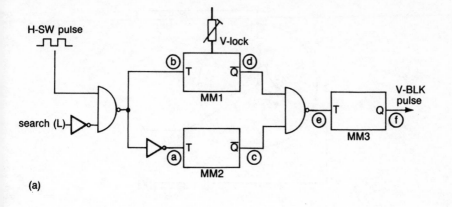

(a)

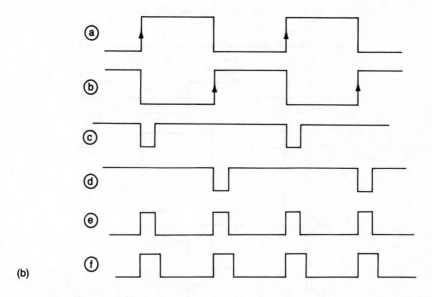

(b)

Figure 4.23 The V-BLK pulse generator and its waveforms: (a) V-BLK generator, (b) waveforms of (a).

the same. Thus:

$$I_1 = i_a - i_b$$

$$= \frac{V_1 - V_2}{2Z} - \frac{V_3 - V_1}{2Z}$$

$$= \frac{2V_1 - V_2 - V_3}{2V}$$

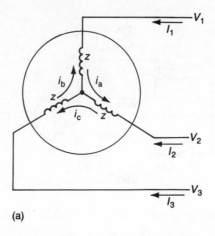

(a)

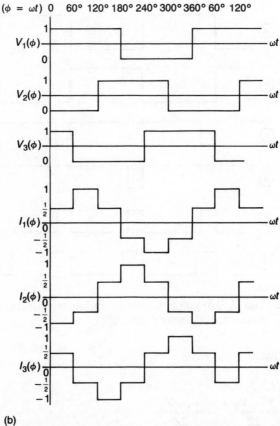

(b)

Figure 4.24 VHS drum and capstan motors: (a) the voltages and currents in three-phase fixed coils, (b) the waveforms of the currents generated by the voltages.

Table 4.2 Three-phase capstan and drum motor voltages

$\phi = \omega t$	0°	60°	120°	180°	240°	300°	360°
$V_1(\phi)$	1	1	1	0	0	0	1
$V_2(\phi)$	0	0	1	1	1	0	0
$V_3(\phi)$	1	0	0	0	1	1	1

For the same reason:

$$I_2 = \frac{2V_2 - V_1 - V_3}{2Z}$$

and

$$I_3 = \frac{2V_3 - V_1 - V_2}{2Z}$$

The values of $V_1(\phi)$, $V_2(\phi)$ and $V_3(\phi)$, shown in Figure 4.24b are listed in Table 4.2.

The values of I_1, I_2 and I_3 can be calculated using the values of $V_1(\phi)$, $V_2(\phi)$ and $V_3(\phi)$ in Table 4.2, resulting in the waveforms of $I_1(\phi)$, $I_2(\phi)$ and $I_3(\phi)$ shown in Figure 4.24b. For example, at $\omega t = 0°$, $V_1(\phi) = 1$, $V_2(\phi) = 0$ and $V_3(\phi) = 1$. With no loss of generality, suppose that $Z = 1$ then:

$$I_1(O) = \frac{2V_1 - V_2 - V_3}{2} = \frac{2-1}{2} = \tfrac{1}{2}$$

$$I_2(O) = \frac{-1-1}{2} = -1$$

$$I_3(O) = \frac{2-1}{2} = \tfrac{1}{2}$$

The levels of $I_1(\phi)$, $I_2(\phi)$, $I_3(\phi)$ at $\omega t = 60°$, 120°, ..., 240°, can be calculated in this manner as shown in Table 4.3. The waveforms of $I_1(\phi)$,

Table 4.3 Three-phase capstan and drum motor currents

$\phi = \omega t$	0°	60°	120°	180°	240°	300°	360°
$I_1(\phi)$	$\tfrac{1}{2}$	1	$\tfrac{1}{2}$	$-\tfrac{1}{2}$	-1	$-\tfrac{1}{2}$	$\tfrac{1}{2}$
$I_2(\phi)$	-1	$-\tfrac{1}{2}$	$\tfrac{1}{2}$	1	$\tfrac{1}{2}$	$-\tfrac{1}{2}$	-1
$I_3(\phi)$	$\tfrac{1}{2}$	$-\tfrac{1}{2}$	-1	$-\tfrac{1}{2}$	$\tfrac{1}{2}$	1	$\tfrac{1}{2}$

$I_2(\phi)$, $I_3(\phi)$ shown in Figure 4.24 are drawn in terms of this table. To sum up, the following conclusions can be drawn.

Three-phase square waves have to be applied to each of the terminals of the motor to obtain three-phase currents in the coils of the motor. These three-phase currents generate a rotary magnetic field and cause a permanent magnet rotator to rotate. Figure 4.25 shows how the three-phase square waves, $V_1(\phi)$, $V_2(\phi)$ and $V_3(\phi)$, are generated at the three terminals of the motor.

The features of the three-phase square waves are:

1. They are three symmetrical square waves.
2. The period of the square waves corresponds to a rotational angle passed through a pair of magnetic poles on the motor.
3. The phase difference between any pair of the three square waves is 120°.

A position signal process, which consists of two or three Hall effect elements can be used to sample the phase signal and generate the three-phase voltage via the differential amplifier and driver. If two sides of the Hall effect element, which has six faces, are excited by a magnetic field and a current, $\mathbf{B}$ and $\mathbf{I}_c$, as shown in Figure 4.25, in which the direction of $\mathbf{B}$ is from top to bottom and the direction of $\mathbf{I}_c$ is from back to front, a Hall voltage (V_H) will be generated on the two sides, left to right, of the Hall element. The polarity and the magnitude of the voltage is dependent on the direction and magnitudes of $\mathbf{B}$ and $\mathbf{I}_c$.

$$V_H = R_H \frac{\mathbf{I}_c \times \mathbf{B}}{d}$$

where R_H is a Hall coefficient.

Thus, changing the direction of either $\mathbf{I}_c$ or $\mathbf{B}$ will result in a change in the polarity of the Hall voltage. The value of the Hall voltage will vary with the intensity of either the current, $\mathbf{I}_c$, or the magnetic field, $\mathbf{B}$. If a Hall element is laid near the motor, the polarity of the Hall voltage will change as the north pole or the south pole of the rotator is closest to the Hall element. This is shown in Figure 4.26.

The three Hall elements, shown in Figure 4.25 (H_1, H_2, H_3) are mounted near the rotator of the motor and the angles between H_1 and H_2, and between H_2 and H_3, are 120°, while the north poles (N) and south poles (S) on the rotator are distributed 180° apart. So the three Hall voltages generated from H_1, H_2 and H_3 each have a phase difference of 120° with respect to each other.

The operating principle of the position signal process detailed in Figure 4.25 is explained in Figure 4.27. In Figure 4.27 N and S indicate the magnetic poles on the rotator and H_1, H_2 and H_3 are the Hall sampling elements. For the purposes of discussion, it can be assumed that the rotator is fixed and the Hall elements are moved to the left, corresponding to a rotation of the rotator to the right.

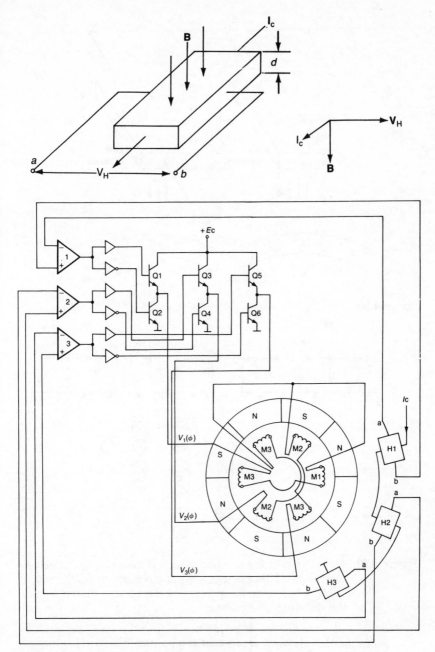

Figure 4.25 Generation of the VHS drum and capstan motor three-phase voltages. Note:

$$\mathbf{V}_{\mathrm{H}} = R_{\mathrm{H}} \times \frac{\mathbf{I}_{\mathrm{c}} \times \mathbf{B}}{d}$$

where R_{H} is a Hall coefficient.

The servo system **127**

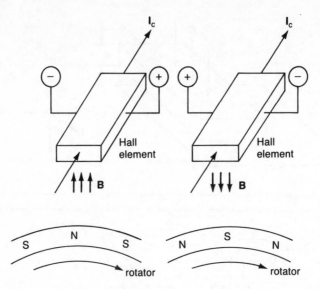

Figure 4.26 The polarity of the Hall voltage varies with moving magnetic field.

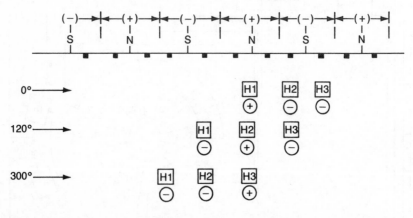

Figure 4.27 Generation of the position signal. Note: $(+)/(-)$ indicates the range of the magnet field at the north or south pole; $\oplus/\ominus$ indicates the polarity of the voltage output from the differential amplifier.

When the Hall elements are located at a position $0°$ and H_1 is located at a section of the north pole, and H_2 and H_3 are located at a section of the south pole, the differential amplifier 1 outputs a positive voltage $(+)$ turning transistor Q1 on and turning transistor Q2 off. In this case, the differential amplifiers 2 and 3 output negative voltages $(-)$ and Q3 and Q5 are turned off while Q4 and Q6 are turned on. The result is that $V_1 = 1$ and $V_2 = V_3 = 0$, i.e. the rotating magnetic field at 60° has been generated as the rotator passes through the 0° position.

Similar conclusions can be drawn for the other positions in which the rotator can pass. For example, when the rotator is in the 120° position H_1 and H_3 are located at a south pole section and H_2 is located at a north pole, so the differential amplifier 2 outputs a positive (+) voltage, while the differential amplifiers 1 and 3 output negative (−) voltages. They turn Q2, Q3 and Q6 on and Q1, Q4 and Q5 off. The result is that $V_1 = 0$, $V_2 = 1$ and $V_3 = 0$, i.e. the rotating magnetic field at 180° is generated. The case when the rotator is in the 300° position can easily be explained by the reader. Therefore, the rotator is driven by the magnetic field which is always in advance of the rotator by 60°.

4.4 The reel servo system

The reel servo system, as mentioned previously, is used to control the tension of the transporting tape, hence it is also called the tension servo system. The reel servo system is relatively simple in VCRs which have no edit function, but is much more complex in machines such as the VO-5850P which have advanced edit functions. A full description of the reel servo system will therefore be based on this model.

The VO-5850P has an enlarged PB speed range so a more complex servo system, controlled by the microcomputer, has been included. The functions of the system are:

(a)   to maintain constant tension when the tape is transported normally;
(b)   to control the change of the tape tension when the direction of the moving tape is reversed; and
(c)   to send a stop signal to prevent damage to the tape and the machine when the tape slackens while in transport.

4.4.1 The reel servo in the VO-5850P

The reel servo system in the VO-5850P comprises three sections which include the mechanism and the circuits and software of the microcomputer. These are shown in Figure 4.28.

The tension of the transporting tape, and the rotational rate of the reel table on the take-up side or supply side are detected and fed to the microcomputer to be used as input instructions $IN5/D_7$, $IN5/D_0$, D_1 and $IN7/D_4$, D_6 (the significance of these inputs is discussed in Chapter 5).

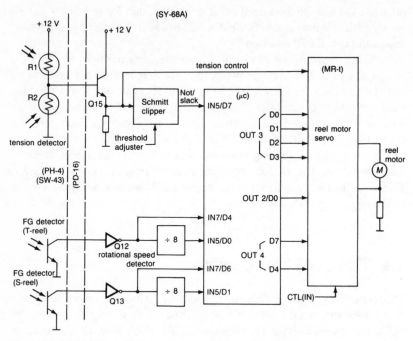

Figure 4.28 The reel servo system in the VO-5850P.

A processing section in the microcomputer generates some output instructions according to these input instructions, and to the condition of the machine. The output instructions, i.e. OUT2/D_0, OUT3/D_0−D_3 and OUT4/D_4, D_7, are combined and are sent to a controlling section as shown in Figure 4.29. The contents of the output instructions are:

D_0 = 1 of OUT2   :   REC
D_0 = 1 of OUT3   :   FWD (reel)
D_1 = 1 of OUT3   :   REV (reel)
D_2 = 1 of OUT3   :   FF/REW (reel)
D_3 = 1 of OUT3   :   STILL (reel)
D_7 = 1 of OUT4   :   pinch solenoid on

In either the normal PB mode, i.e. D_0 = 1 of OUT3 and D_7 = 1 of OUT4, or the REC mode, i.e. D_0 = 1 of OUT3 and D_0 = 1 of OUT2, IC126-4 is high, i.e. a FWD signal is fed to the reel motor servo and inhibits the stop signal. In the REV mode, i.e. D_1 = 1 of OUT3, a high logic level signal known as the REV signal, is fed to the control system and also inhibits the stop signal. The case of the FF/REW mode is the same, D_2 = 1 of OUT3, i.e. the FF/REW signal, is present and the stop signal is inhibited. If the machine is not in the FF, REW or FWD, REV mode, i.e. D_2 = D_1 = D_0 = 0 of OUT3, the stop mode is present. This means that IC124-10 is high.

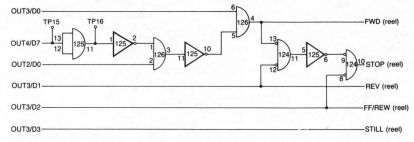

Figure 4.29 Input/output condition control.

When the SEARCH dial is turned to the × 0 position, $D_3 = 1$ of OUT3, i.e. a still signal is generated. The difference between the stop signal and the still signal is that the stop signal can be locked with the FWD, REV and FF/REW signals, but the still signal is independent of these modes. This indicates that all the modes, the FWD, the REV and the FF/REW modes, are released if the stop mode is set up.

In addition, a special output instruction, $D_4 = 1$ of OUT4, is referred to as a × 10 search signal. This indicates that a small cassette, KCS type, is used and that the machine is in the search mode. In this case, the highest tape speed which can be achieved is × 10 normal tape speed. This is referred to as a U-scan mode.

A control section is located on the reel motor servo. This section controls the rotational rate of the reel motor in terms of the instructions from the microcomputer, with the detected tape tension and the CTL signal indicating the tape speed. These circuits are described below.

4.4.2 The tape tension detector

The tape tension detector converts the detected tape tension into an electrical signal by passing the light from a light emitting diode (LED) through the slit in a shutter which is connected to the tension detect arm. The shutter is located on either the take-up side or supply side in the FWD or REV modes, respectively, and projects onto a divider comprising a set of light dependent resistors (LDR), i.e. CdS element, as shown in Figure 4.30.

When the tape tension changes, the tension arm is moved and the shutter is turned around a shaft so a ray of light on the CdS resistors is also moved. The conductors (+), (−) and (s) are used to connect the CdS resistors and the external circuit, the resistor from (+) to (s) in the lower left side is known as R1 and the (s) to (−) resistor in the upper right side is referred to as R2. The two LDRs are constructed to form an interleaved comb so that the movement of a ray of light causes one LDR to increase in resistance while the other decreases. This provides a relationship between the tape tension and

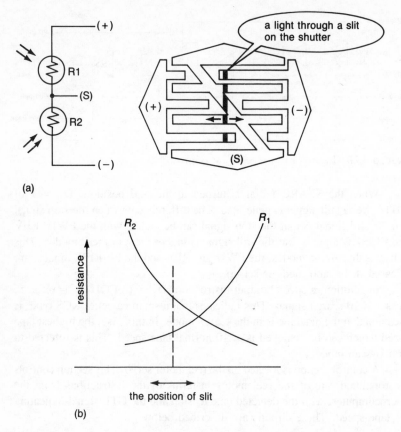

(a)

(b)

Figure 4.30 Tape tension detector: (a) CdS elements, (b) resistance characteristic of CdS elements.

the divided voltage formed by R1 and R2. This divided voltage is referred to as the CdS TENREG (tension regulator) signal. The TENREG signal is divided into two paths after being buffered by Q15. One path feeds the signal directly to the reel motor servo and is used as one of the input signals of the reel motor servo. The other is fed to IC110-3 of a comparator, IC110. The output at TP18 is high as the voltage at IC110-3 is greater than the comparison voltage, and is low in the reverse case. By adjusting RV3 in Figure 4.31 the variable range of the tape tension can be limited. This is known as the window of the detected signal.

If the tape tension varies over the 'window', the level at TP18 will be reversed and used as a detected signal to be input into the microcomputer, i.e. IN5/D_7, to slacken the tape. This is a sort of auto-stop signal. Figure 4.31 shows part of this circuit. Two photoelectric detectors, shown in Figure 4.28,

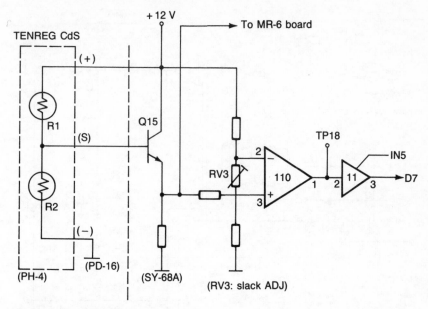

Figure 4.31 TENREG circuit.

are used to detect the reel speed on the take-up side and the supply side. In the FWD mode, the detected reel speed on the take-up side is known as the T-FG signal as the tape is taken up by the take-up reel. The detected reel speed on the supply side is referred to as the S-FG signal as the tape is taken up by the supply reel. Either the T-FG or the S-FG signal is first divided by eight before being fed to the microcomputer, i.e. $IN5/D_0$ or D_1.

4.4.3 The reel motor servo

The circuit for the reel motor servo is shown in Figure 4.32. The reel motor is driven by current supplied from a power transistor, Q1, and a drive transistor. The rotational speed of the motor is dependent on a driving signal applied to the base of Q25. The driving signal differs in different modes. There are four switches, S1, S2, S3 and S4 (shown in Figure 4.32), which are controlled separately by the × 10 SEARCH signal, and the FWD, REV signals.

In the × 10 SEARCH mode only, a high logic level at IC4-2 is applied to S1 and turns it on, while a low logic level at IC4-10 is applied to S2 and turns it off because $D_4 = 1$ of OUT4 at this point. The opposite situation means that it is not in the × 10 SEARCH mode, but it is available in the FWD, REV, FF/REW, stop or still modes. In these modes both switches, S1 and S2 are switched over, i.e. S2 is turned on and S1 is turned off because

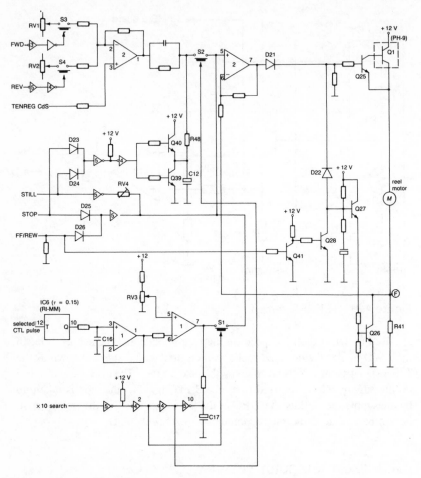

Figure 4.32 The reel motor servo.

$D_4 = 0$ of OUT4 in this case. This can be summarized as:

1. The switch S1 is turned on by the FWD signal in the FWD mode.
2. The switch S4 is turned on by the REV signal in the REV mode.

Normal playback

In the playback modes, including normal or search modes but excluding the still mode, the rotational speed of the motor is controlled by the TENREG CdS signal, i.e. a divided voltage from two LDRs, R1 and R2, when the tape tension is smaller than normal, i.e. a slack tape, the shutter is moved to the right, R2 increases and R1 decreases so the divided voltage increases. This

134 Video recorders

TENREG signal is amplified and used to drive the reel motor. The result is that the motor is driven faster.

An idler, referred to as the FF/REW idler, is connected to the reel motor by a transmission belt and is attracted against the take-up reel by a FWD idler solenoid in the FWD mode, or against the supply reel by a REV idler solenoid in the REV mode. No matter which mode is selected, the take-up reel, or the supply reel, will be driven faster in order to take up the slack tape. The result is to increase the tape tension until normal tension is reached. The discussion is the same for the case when the tension is greater than normal.

It should be noted that because the tape tension in the FWD and REV modes is different, two offset voltages in the FWD and REV modes, respectively, are used and are adjusted by RV1 and RV2 using switches S4 and S3 to select the appropriate offset.

Stop or still modes

In the stop or still modes, Q39 and Q40 are switched on by the stop or still signals, so the output voltage at IC2-1 is decreased because R48 is grounded. This is used for taking up the tape which is transported at $\frac{1}{15}$ of normal speed by the capstan. To prevent slackening of the tape, the reel speed must be decremented a short time after that of the capstan. The delay circuit consisting of C12 is used for this purpose.

FF or REW

In the FF or REW modes the FF/REW signal is split into two control signals. The first is the same as the stop signal, it shorts out the level at IC2-5 after being inverted by IC5, making the output from the operational amplifier, IC2, zero. The other control signal is used to turn Q41 on and turn Q28 off. This feeds +12 V to the base of the drive transistor, Q25, through diode D22. Meanwhile, Q27 and Q26 are turned on, shorting out the feedback resistor R41. The output drives the reel motor to maximum speed.

Tension control in the FF/REW modes is similar to that in the PB mode described earlier. The FF/REW idler is pulled against the take-up reel when the FWD mode is selected and the FWD solenoid is on. The idler is pulled against the supply reel, when the REW solenoid is on, in the REV mode.

Search

When a cassette of the KCS type is used, such as the KCS-20 or the KCS-10 made by the Sony Company, and the search dial is turned to the $x(\pm 5)$ position, the $\times 10$ SEARCH mode is set up. As mentioned earlier, the $\times 10$ SEARCH signal, i.e. $D_4 = 1$ of OUT4, turns the switch S1 on and S2 off. This means that the signal applied to IC2-5 originates from IC1-7 instead of IC2-5. In this mode, the pinch solenoid is turned off and the tape is driven by the reel motor instead of the capstan motor. The selected CTL driving pulse, from the servo system, is applied to a retriggerable monostable, IC6. When the period of the selected CTL signal is less than the time constant, $\tau = 0.1$ s, of the retriggerable monostable, the output at IC6-10, i.e. the Q terminal, is at a high level. It is applied to a capacitor, C16, and then compared with a reference voltage, adjusted by RV3, through an operational amplifier, IC1. The difference voltage at IC1-7 is applied to IC2-5, Q25 and Q1 and controls the reel motor.

4.4.4 Summary of the reel servo system

During forward tape transport in the modes including the REC mode, normal PB mode, the search (forward) mode and the unthreading process (but not including the FF mode), the tape tension servo system operates in the following manner.

At the supply side the S-tension regulator arm is pulled to the right by the spring when tape tension is low, so the felt band is pulled tightly around the supply reel, as shown in Figure 4.33. The result is that the speed of the supply reel is reduced and the tension of the tape on the supply side is automatically increased. The reverse happens when the tape tension is high.

A SKEW adjustment is fitted on the S-tension regulator arm to allow manual control of the tape tension in the PB mode. It is necessary to use this adjustment if the length of the tape in the PB mode is different to that in the REC mode. This may occur when conditions such as temperature, humidity and tape tension are not the same in both modes and would result in the length of the video track being different. This would produce a discontinuity in the PB video signal at the switching point, and if the deviations were large enough a skew bar would appear at the top or the bottom of the screen because the line synchronization could not be locked. This may be rectified by adjusting the skew knob on the front panel of the machine to ensure that the length of the line in the PB mode matches that of the track in the REC mode.

At the take-up side, the tension is achieved by both mechanical and electrical means. This is shown in Figure 4.34. When tension is too high the shutter is moved to the left and the divided voltage from the CdS detector is reduced. The reel motor rotates, followed by the FF/REW idler, and the take-up reel rotates more slowly, resulting in reduced tension as the tape is taken up more slowly.

While the tape is being transported in the reverse direction, including the search (reverse) mode at any speed, but excluding the REW mode, the tension servo operates as described below. At the supply side, which is in fact the take-up reel since the direction is reversed, when the tension is low the T-tension regulator arm is pulled to the left and a reel-brake is pulled against

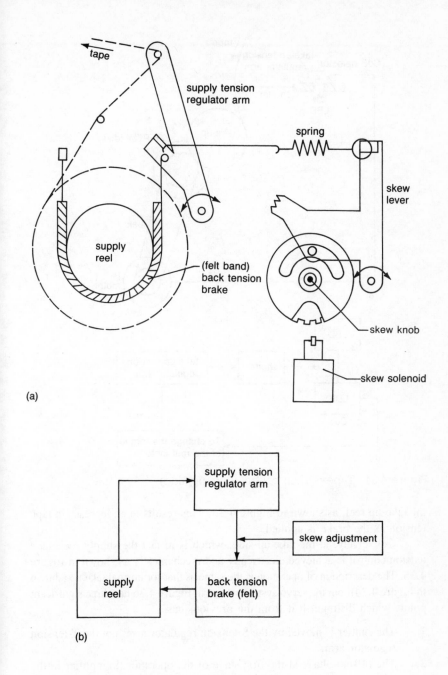

(a)

(b)

Figure 4.33 Tension control (supply side).

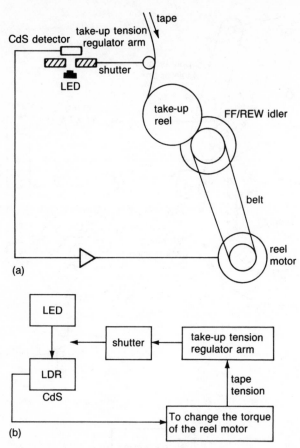

Figure 4.34 Tape tension control (take-up side).

the take-up reel, as shown in Figure 4.35. This results in an increase in tape tension as the brake is applied.

Conversely, at the take-up side, which is in fact the supply reel, tape tension control is achieved electrically and mechanically as shown in Figure 4.36. The description of operation is the same as that for normal take up (shown in Figure 4.34), but the servo system shown in Figure 4.36 has three significant points which distinguish it from the previous case.

1. The shutter is moved by the S-tension regulator arm, not the T-tension regulator arm.
2. The offset voltage at the first stage of the operational amplifier in the reel servo circuit is supplied by RV2 not RV1 as shown in Figure 4.32.
3. The FF/REW idler is pulled against the supply reel not the take-up reel.

In the FF or REW modes the reel motor is driven without any feedback loop.

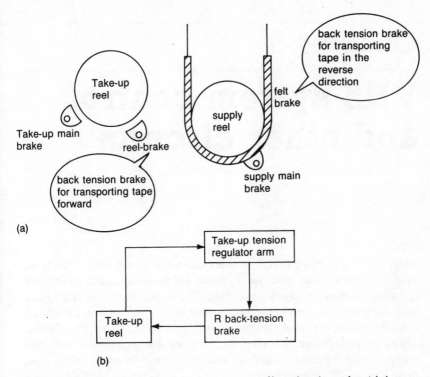

Figure 4.35   Tension control in the reverse direction (supply side).

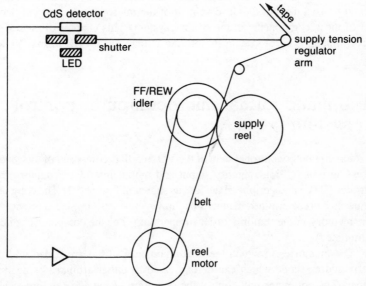

Figure 4.36   Tape tension control in the reverse direction (take-up side).

Chapter 5

VHS system control and other circuits

System control within the VCR includes control of all the logical functions, input signal instructions from the key panel, and the signals detected from the operating states and modes of the machine. The output signals from the system control are used to control mechanisms such as the reel motor, the threading or loading motor, the cassette compartment or front loading motor, the solenoids driving the pinch roller, the brake, etc. They are also used to control circuits such as the video and audio channels, and servo system for switching the mode from REC to PB or reverse.

In early VCR models the control system consisted of a number of logic ICs and transistors, but the development of the microprocessor has replaced most of these components in the control system. This chapter discusses the microcomputer control system in both the U-matic and VHS format VCRs.

5.1 U-matic format microcomputer control system

The microcomputer control system of the VO-5850P is composed of the central processing unit (CPU), memory, input and output interface, counter/timer controller (CTC) and peripheral devices as shown in Figure 5.1. The computer operates in a predetermined manner by following instructions placed previously in the memory in the required order of operation (i.e. the computer has been programmed).

The memories in which the instructions are stored are coded with an orderly address code which can be expressed as either a binary or decimal number. The computer will control the operation of the VCR automatically provided the CPU follows the order of the address code to read the instructions

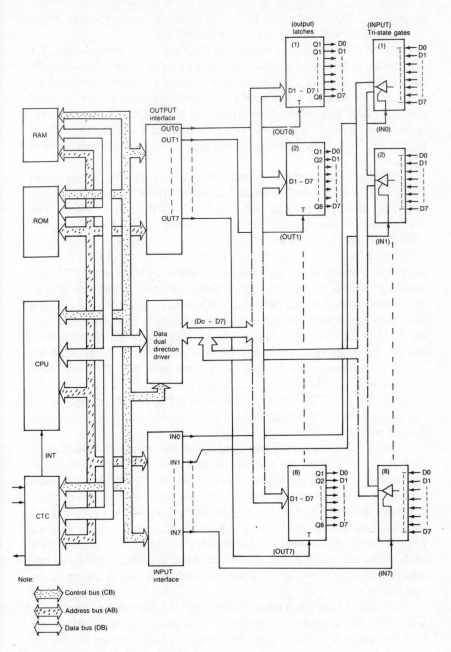

Figure 5.1 The microcomputer control system.

and analyse them. The control signals are then sent to the appropriate memories, Input/Output (I/O) interfaces and CTC to execute the basic operations which were originally indicated by the stored instructions.

If the CPU receives an instruction from one of the input devices, for example D_2 of IN4 or D_6 of IN4, etc., or receives an interrupt request from the CTC, the executing program can be transferred to another segment of the program. This is known as a subroutine or interrupt service procedure, in which one, or a few, new output instructions will be generated and sent to the appropriate output devices in the VCR, for example D_4 of OUT5 or D_2 of OUT7, etc. This enables interactive operation of the microcomputer control system.

The operating principles of the microcomputer control system shown in Figure 5.1 will be based on model VO-5850P. There are three buses: data bus (DB), address bus (AB) and control bus (CB) in this system. The microcomputer control system can detect 64 inputs, D_0-D_7 of IN0, D_0-D_7 of IN1, etc., to D_0-D_7 of IN7. It can also output 64 states to control up to 64 functions in the VCR, i.e. D_0-D_7 of OUT1 to D_0-D_7 of OUT7. These input or output data on the data bus can be delivered between the CPU and peripheral circuits through the dual direction data drivers and data latches. The random access memory (RAM) can be used as a register to store the input and output data. This means that the output data must be first transferred from the RAM to the CPU, then output from the CPU to the peripheral devices and vice versa. Instructions to carry this out are previously placed in ROM.

Three addresses, A_0, A_1 and A_2, are sent from the CPU to the input or output interface where the three addresses generate eight selecting signals, i.e. IN0−IN7 or OUT0−OUT7. Thus one of the eight input or eight output latches can be selected and eight bits of data, D_0-D_7, can be passed through the selected gate or latch.

The CPU can also exchange signals with the peripheral devices using the CTC, in which there are four independent channels. These channels can be used for the timer or the counter. For example, in VO-5850P, the function of channel 2 is to generate the search frequency and it operates as a counter. The input signal of the channel is the clock frequency which is derived from the crystal oscillator in the VCR. The period of the search frequency signal, which is output from channel 2, is controlled by four data bits from the position cords of the search dial on the front panel. More accurately, three data bits determine the search frequency, while the fourth bit is used to determine the tape direction.

Channel 3 is used as a timer. Every 2.08 ms an interrupt request is sent to the CPU. The interrupt service routine will automatically be executed when the CPU responds to the request. The contents of the interrupt routine are instructions to scan the keyboard and the timer display. These are shown separately in Figures 5.2 and 5.3 and are explained briefly below.

The interrupt routine produces a rising edge at D_4 of OUT0 every

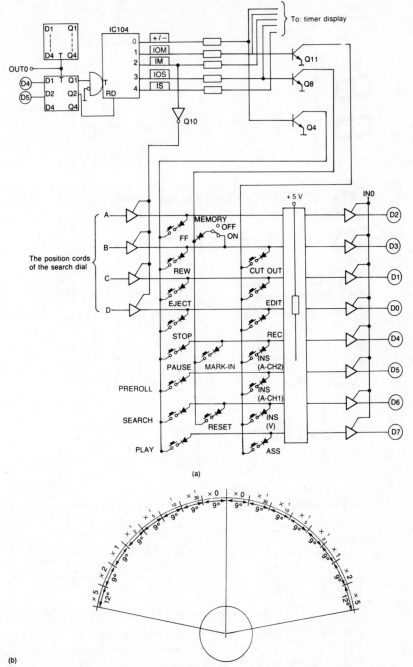

Figure 5.2 The search key instructions: (a) scanning circuit of the key matrix, (b) the position of the search dial.

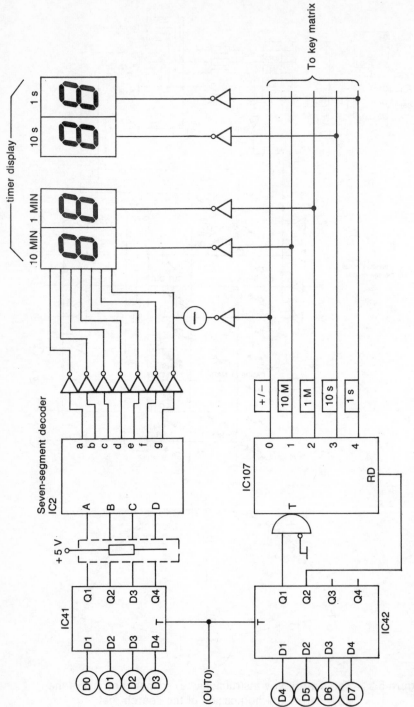

Figure 5.3   Key matrix and timer display scan.

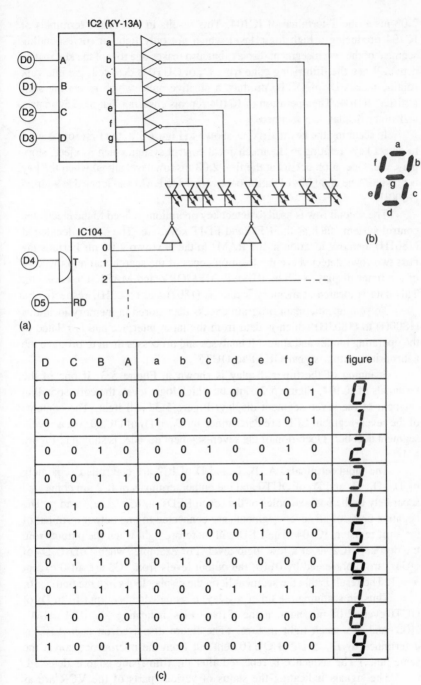

D	C	B	A	a	b	c	d	e	f	g	figure
0	0	0	0	0	0	0	0	0	0	1	0
0	0	0	1	1	0	0	0	0	0	1	1
0	0	1	0	0	0	1	0	0	1	0	2
0	0	1	1	0	0	0	0	1	1	0	3
0	1	0	0	1	0	0	1	1	0	0	4
0	1	0	1	0	1	0	0	1	0	0	5
0	1	1	0	0	1	0	0	0	0	0	6
0	1	1	1	0	0	0	1	1	1	1	7
1	0	0	0	0	0	0	0	0	0	0	8
1	0	0	1	0	0	0	0	1	0	0	9

(c)

Figure 5.4 Operation of the seven-segment decoder: (a) circuit, (b) LED, (c) the truth table.

2.08 ms at the T-terminal of IC104. This results in one of the terminals of IC104 producing a high logic level which not only lights a corresponding segment of the seven-segment display, but also makes one row of the key matrix active. When the fifth rising edge from D_4 of OUT0 is detected, the interrupt routine makes D_5 of OUT0 produce a positive pulse which resets the shift register, IC104. The operation of IC104 repeats ensuring that the key matrix and timer display are scanned.

In scanning the key matrix, as shown in Figure 5.2, the first row is used to detect keys relating to the mechanical control system, such as eject, stop, play, etc. Operation occurs at the first 2.08 ms low level signal when the key is pressed. The eight bits of data are transferred to RAM and located at address (1800H).

The second row is used to detect key operations related to the electronic control system, such as the REC and EDIT keys, etc. The data is located at (1801H) memory location in the RAM. In the next two 2.08 ms periods the next two rows detected are the location codes of the search dial and the keys of the timer display, such as RESET, MEMORY and MARK-IN keys, etc. This data is located at memory locations (1802) and (1803H), respectively.

To sum up, the main program checks data stored in memory locations (1800H) to (1803H) which is data from the input interface and is related to the operating modes and states. It analyses and processes them to produce the appropriate output signal OUT1 to OUT7.

Scanning of the timer display is shown in Figure 5.3. If one of the terminals of IC104, Figure 5.2, outputs a high logic level, the corresponding segment of the seven-segment display, Figure 5.3, will light. The contents of the seven-segment LED are determined by D_0 to D_3 of OUT0, via a seven-segment decoder. Operation of the seven-segment decoder is shown in Figure 5.4.

The input terminals, A, B, C and D of IC2 are controlled separately by D_0, D_1, D_2 and D_3 of OUT0 and the output terminals of IC2 are connected separately to the positive poles of the seven LEDs, a, b, c, d, e, f and g. The negative poles of all seven segments are connected to one output terminal of the shift register, IC104. The LED will therefore light when the output from the decoder, IC2, is at a low logic level. For example, when $D_3D_2D_1D_0$ of OUT0 are represented by 0110, the output levels from IC2 are all 0 except $b = 1$. The result is that the segment lit on the display looks like the number 6.

Thus, in scanning the timer display, a rising edge is output from D_4 of OUT0 every 2.08 ms and then one of five LEDs, indicating '-', '10M', '1M', '10S' and '1S' each light in turn. (The figure displayed on each LED is determined by D_0 to D_3 of OUT0 sent out from the microprocessor at the same time.) The sequence is repeated after the fifth rising edge is detected.

The signals indicating the status of various parts of the VCR are as follows:

1. Check sensor signals at the beginning and end of the tape.
2. Condition signals from photoelectric switches in the threading system.
3. Fault signals, e.g. high humidity, or drum starting faults.
4. Fault signals such as tape slackness.

These signals are input through the input interface, IN2 to IN6, when they are detected by the peripheral circuits.

The output signals from the microprocessor control system, as mentioned above, are used mainly to control the mechanical system, the electronic circuits and indicators, for the operating modes of the machine.

5.2 VHS system control

The system control circuits consist of one or two monolithic IC processor chips. The block diagram of the VHS system control is shown in Figure 5.5. The peripheral circuits of the monolithic processors comprise two sections, i.e. input and output sections. The information input into the monolithic processor is derived from the following units or circuits:

(a)   the function key panel;
(b)   a wired or infra-red remote controller;

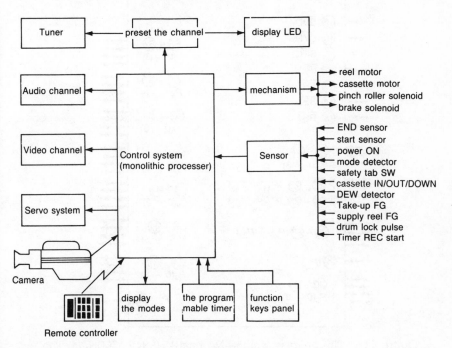

Figure 5.5   VHS system control block diagram.

(c) the programmable timer;

(d) the mode or condition detectors (or sensors); and

(e) the camera

Control signals from the monolithic processor are sent to the following units or circuits:

(a) the video and audio channels;

(b) the servo system;

(c) the mechanism, including the solenoids, reel and cassette motors, etc.; and

(d) the display unit for displaying the time and operating modes, etc.

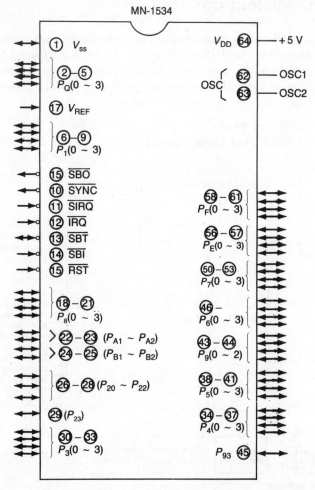

Figure 5.6   The processor in the National VHS NV-370EN. (a) The pins of the monolithic processor.

Structure of monolithic processor

The block diagram and pin connections for the National VHS type NV-370EN monolithic processor are shown in Figure 5.6. The monolithic processor (MCU) generally consists of the CPU, memory (ROM and RAM), I/O interfaces, CTC and clock in a single chip. There are two types of I/O interfaces, i.e. four-bit parallel ports and a serial port, but no CTC in the chip shown in Figure 5.6.

The MCU is a 64-pin dual in-line package, and with the exception of four pins, V_{SS}, V_{DD}, OSC1 and OSC2, can be divided into two main types. These are pins relating to the data bus and the control bus. There are thirteen sets of pins for the four-bit parallel ports and a set for the serial port in the data bus. Each set of pins for the four-bit parallel ports are called, respectively, $P_1(0-3)$, $P_2(0-3)$, ..., $P_9(0-3)$, $P_A(0-3)$, $P_B(0-3)$, $P_E(0-3)$ and $P_F(0-3)$. They are used as both the data bus and address lines at different times according to the MCU instructions.

The input and output serial bus ports are known as SBI and SBO, respectively, and are effective when a low logic level is sent to the SBT (toggle)

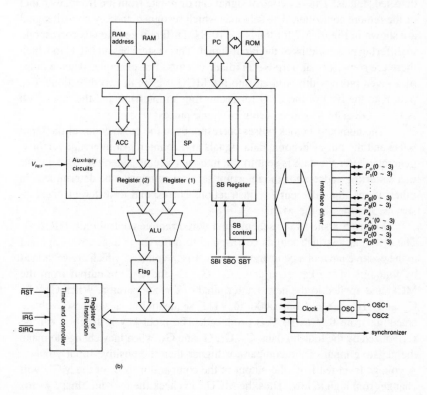

(b)

Figure 5.6 (contd) (b) The block diagram of the monolithic processor.

terminal. They are used to send the function indications in the NV-370EN. The control bus shown in Figure 5.6 can be described further.

$\overline{RST}$ the chip is reset when a logic low level is sent to this terminal

$\overline{IRQ}$ and $\overline{SIRQ}$ non-maskable and maskable interrupt request

V_{REF} a reference voltage provided for special purpose auxiliary circuits inside the MCU

$\overline{SYNC}$ the synchronizing signal from the internal clock is output at this terminal

$\overline{SBI}$, $\overline{SBO}$ and $\overline{SBT}$ have been explained above. $\overline{SIRQ}$ and $\overline{SYNC}$ are not used in NV-370EN. Use of the MCU pins in the NV-370EN model are shown in Table 5.1.

Key control signals and detected signals

The input signals to the MCU can be divided into key control signals and detected signals. The key control signal can originate from the front key panel or the remote controller. The schemes which recognize the key control signal are shown in Figure 5.7 for the NV-370EN. Different key signals correspond to differing pulse widths of the input signal. This system has an MCU in which there are some special purpose auxiliary circuits. For example, when a pulse of a given pulse width is input into the MCU it goes in two directions, one path is to the IRQ of the CPU for requesting an interrupt and the other path is to an OR gate to gate a string of clock pulses.

The number of clock pulses corresponds to the pulse width of the input pulse and the pulses become data signals at a counter. An interrupt vector is generated after the data is sent to the interrupt vector generator. This means that a starting address for the interrupt service routine can be determined. In other words, the different key instructions correspond to different interrupt service routines in the MCU.

Figure 5.8 shows the scheme in a wired remote control unit, HR7600. The different key instructions correspond to different voltage levels applied to the positive input of a comparator. Comparison voltages, which are generated by four bits of the key scan data, G_0, G_1, G_2 and G_3, are output from the MCU and applied to the negative terminal of the comparator. When G_0, G_1, G_2 and G_3 change from 0000 to 1111 sequentially, a staircase voltage changing from 0 V to 10 V is produced. The input key instruction can be recognized by the four-bit data, G_0, G_1, G_2 and G_3, when they change to make the negative input of the comparator higher than the positive input voltage. A voltage level fed from the output of the comparator to B_2 of the MCU will change from high to low. Thus the MCU can check the four-bit binary words on G_0 to G_3 as the level of the input port B_2 changes from high to low, and hence recognize which key has been pressed.

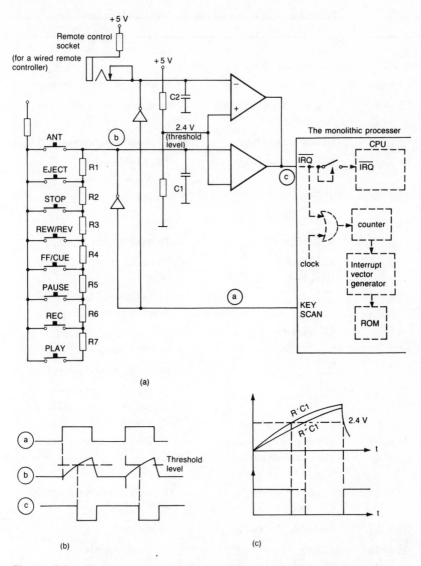

Figure 5.7 Key input control in the VHS NV-370EN: (a) the key input circuit in NV-370EN, (b) the waveforms of points a, b and c in the circuit, (c) the different charging resistor corresponds to the different pulse width.

VHS system control and other circuits **151**

Table 5.1 Use of the pins in the NV-370EN model

Pin	Symbols	Use
1 and 64	V_{SS} and V_{DD}	Ground and supply power
2 to 4	Cassette IN/OUT/DOWN (P_{ϕ^0} to P_{ϕ^2})	Condition signals IN
5	Camera pulse (P_{ϕ^3})	To select camera input mode
6 to 9	Front load and load ($P_{\phi^{10}}$ to $P_{\phi^{13}}$)	To drive the motors clockwise or anticlockwise
10	$\overline{SYNC}$	No use
11	$\overline{SIRQ}$	No use
12	$\overline{INT}$ (IRQ)	To input the key signal
58 to 59	Key scan (P_{F1}, P_{F2})	To output the key scan pulse
13	$\overline{SBT}$	To transfer the function indicated signal
14 to 15	$\overline{SBI}/\overline{SBO}$	To transfer the function indicated signal
16	$\overline{RST}$	Reset the chip on power-up
17	V_{REF}	To provide a reference voltage to special purpose auxiliary circuits in the MCU
18 to 21	H-SW/reel sensor/supply photo/take-up photo (P_{80} to P_{83})	To input the detected signals
22 to 23	Power ON/DEW (H) (P_{A1} to P_{A2})	To input the detected signals
24 to 25	6H/index (P_{B1} to P_{B2})	No use
26 to 28	$S_1/S_2/S_3$ (P_{20} to P_{22})	The mode signals

Figure 5.9 shows the scheme adopted in the NV-788; this is the same as that in VO-5850P. The key scan signals are output from $PO_{(0-3)}$. The key scan signal is fed to $PD_{(0-3)}$ and are stored, respectively, in adjacent memory locations, for example:

1. The key signal, FF, REW, EJECT or STOP, sampled by the pulse PO_0 is stored in the N memory location.

Table 5.1 (contd)

Pin	Symbols	Use
29	Safety tab (P_{23})	To prevent misrecording
30 to 33	Memory SW/clear SW/timer REC/timer set (P_{30} to P_{33})	For the timer display of time
34 to 37	CYL ON/STILL/$\times$ 4 (P_{40} to P_{43})	To the servo system (P_{43} is not used)
38 to 41	Cap ON/REV/$\times$ 4/$\times$ 8 (P_{50} to P_{53})	To the servo system
42 to 44	Television/delayed REC/delayed AREC (P_{90} to P_{92})	To the video and audio channels
45	Serial clock (P_{93})	For the serial data
46 to 48	(P_{60} to P_{62})	No use
49	During unload (P_{63})	To the servo system
50	Search (P_{70})	To servo system
51 to 52	EE(L)/Audio muting (P_{71} to P_{72})	To the channel and servo system
53	Sensor LED (P_{73})	To light the LED to detect the tape end or beginning
54	Load V up (P_{EO})	To control the loading motor
55 to 56	Audio DUB/REC (P_{E1} to P_{E2})	To the channel
57	Power ON (P_{E3})	To supply power to the whole machine
60 to 61	(CUE/REV/frame advance) (P_{F2} to P_{F3})	To the servo system
62 to 63	OSC1, 2	

2. The key signal, PAUSE, sampled by the pulse PO_1 is stored in the $(N+1)$ memory location.
3. The key signal, REV, PLAY, A DUB or REC, sampled by the pulse PO_2 is stored in the $(N+2)$ memory location.
4. The key signal, SLOW or STILL/ADV, sampled by the pulse PO_3 is stored in the $(N+3)$ memory location.

VHS system control and other circuits **153**

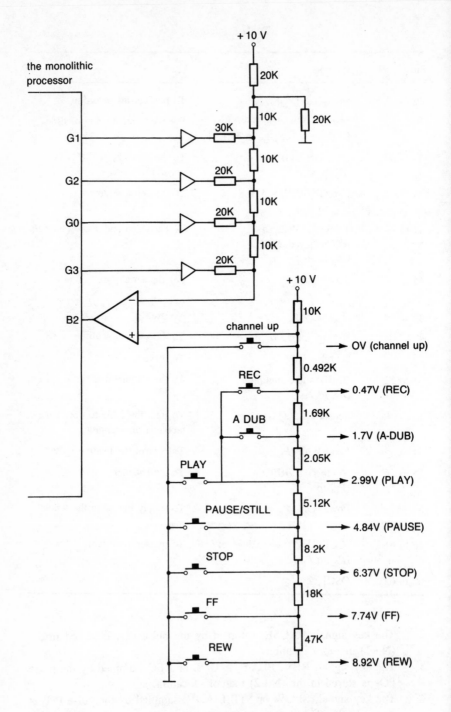

Figure 5.8 Key control in the HR7600 wired remote control.

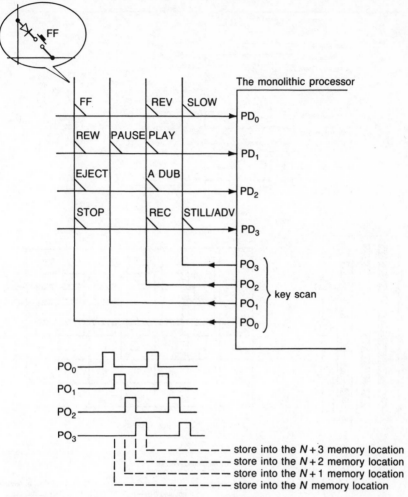

Figure 5.9   Input key control in the NV-788.

Thus the MCU can recognize which key has been pressed if a high level in the four-bit data is picked out from the corresponding memory location. The detected signal from the appropriate sensor inside the machine may be placed in two categories, i.e. the signal detecting the operating condition and the auto-stop signal which prevents damage to the machine. These are shown in Figure 5.10.

The VTR switch is on the front panel of the machine and the MCU has to be powered up with the VTR switch turned to ON for the MCU to perform the appropriate program. The MCU is reset and performs the initialization program when the machine is powered up and a low level is applied to the $\overline{\text{RST}}$ terminal of the MCU.

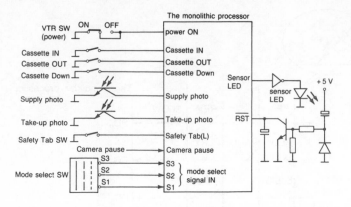

(a)

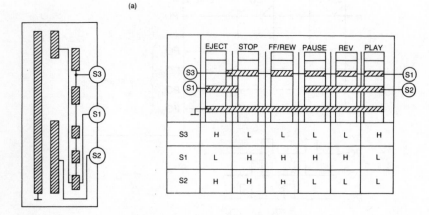

(b)

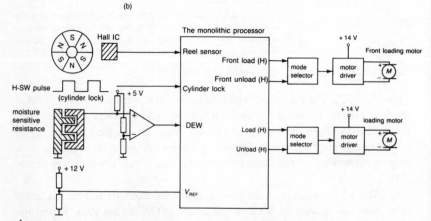

Figure 5.10 Operating condition detection: (a) the input signals to the MCU, (b) the mode detector, (c) the input and output signal of the MCU.

Table 5.2 Control of loading and front loading motors: H = logic level high
L = logic level low

	Eject	Loading and unloading	Cassette down
Cassette IN SW	H	L	H
Cassette OUT SW	L	H	H
Cassette DOWN SW	H	H	L

Three switches, cassette IN, cassette OUT and cassette DOWN all act in association with each other. They are placed so as to generate the data shown in Table 5.2 and to control the loading motor and the front-loading motor.

Two photoconductors are arranged on both the supply and take-up sides to detect the tape end and tape start signals, i.e. a low level. This can be achieved by transmitting light between a sensor and LED through the transparent tape at the start and end of the tape. The safety tab switch is turned on and a low level is applied to the MCU if the tab on the cassette has been removed. This means that the cassette can be used for playback only and will not record.

Three mode select switches, S1, S2 and S3, are selected by the loading motor via linkages. These act in association with each other to send a set of data to the MCU. The program in the MCU performs according to the input key signal and the mode data, as shown in Figure 5.10b.

The special purpose auxiliary circuit in the MCU needs a reference voltage which is provided by a voltage divider circuit and applied to the V_{REF} terminal of the MCU. An H-SW pulse is applied to the cylinder lock terminal of the MCU when the drum servo has been locked. This pulse has to become a high or low logic level before it is sent to the CPU. This is achieved using, for example, an RE-MM in the CPU.

Both the reel sensor terminal and the DEW terminal on the MCU receive the auto-stop signal when either the take-up table is rotating slowly and the tape is loose, or when the moisture in the drum region is higher than normal causing the moisture sensor resistance to fall to zero.

Output signals

The output signals of the MCU may be placed in the following categories:

1. The signals controlling the video and audio channels is comprised mainly of the signal selecting the input signal to the channel, and the signal selecting REC or playback mode of the channel, such as V (video), delayed REC, AREC, EE and audio muting, etc.

2. The servo control signal consists of the signal which causes the drum and capstan motors to rotate or stop, and the signal which changes the speed and direction of rotation of the capstan motor, such as CYL ON, CAP ON and REV, $\times 2$, $\times 4$, $\times 8$, frame advance, etc.

3. The signals controlling the front-loading motor and the loading motor drive in either the clockwise or anticlockwise direction as shown in Figure 5.10c.

4. The serial data from the SBO terminal of the MCU is used to transfer the function indicate signal. The serial clock is output from the SBT terminal. This data is decoded to a set of segment pulses, as shown in Figure 5.11.

The operating principle of the function indicator is similar to the timer display unit in the VO-5850P. There is an oscillator in an IC of type MN1450BVF2. The grid pulse generated by the oscillator is output from C_{00}, C_{01}, $\ldots$, C_{07} sequentially to light a corresponding segment of the seven-segment valve. The contents of the seven-segment valve are determined by the segment pulse at any given time. For example, the values of 1G, 2G, $\ldots$, 6G in the function generator are shown in Figure 5.12d. The grid pulse is applied to the plate of each valve in turn and the segment pulse controls seven grids of the valve. The configurations of 1G, 2G, $\ldots$, 6G are shown, respectively, in Figure 5.12a.

The operation of the key scan and display scan relating to the timer, as shown in Figure 5.13, is the same as that shown in Figure 5.11 except that a seven-segment LED is used instead of a seven-segment valve.

Running through the program

The operation of the program in the MCU can be outlined simply. The program usually consists of two sections, the main program and the interrupt service routine. The contents of the main program include the following parts:

1. An initialization program for the whole system.
2. A subroutine to detect the mode of the machine at power-up.
3. The program to detect the mode or condition of the machine.
4. The display subroutine.
5. The program displaying the status of the fault and protection mechanism.
6. The delay subroutine.

The flow chart for the main program is shown in Figure 5.14.

After initialization, the subroutine to detect the mode of the machine is first called followed by enable interrupt. The second step of the main program checks the following conditions:

1. Has a fault occurred?
2. Is the machine on power-up?
3. Is the cassette inserted?

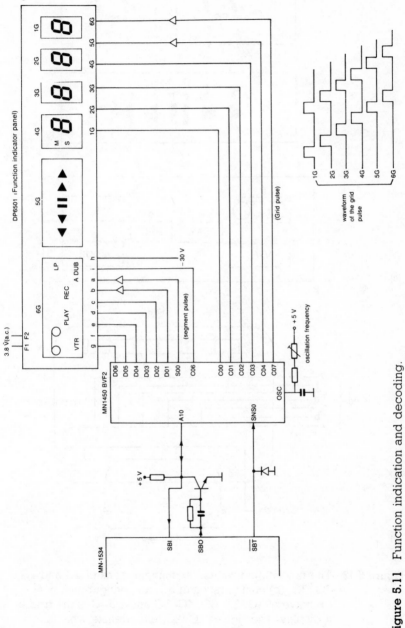

Figure 5.11 Function indication and decoding.

VHS system control and other circuits **159**

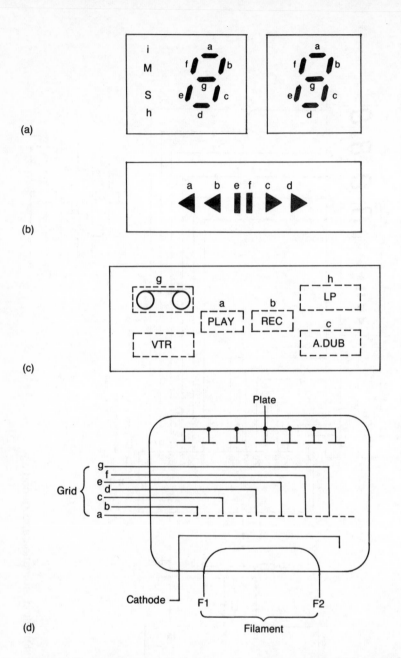

(a)

(b)

(c)

(d)

Figure 5.12 The function indicators: (a) configurations of 4G and 1G, 2G, 3G, (b) configuration of 5G, (c) configuration of 6G, (d) valve of 1G, 2G, 3G, 4G, 5G and 6G. G is the grid of a display. The signals 1G, 2G, etc., denote what is displayed.

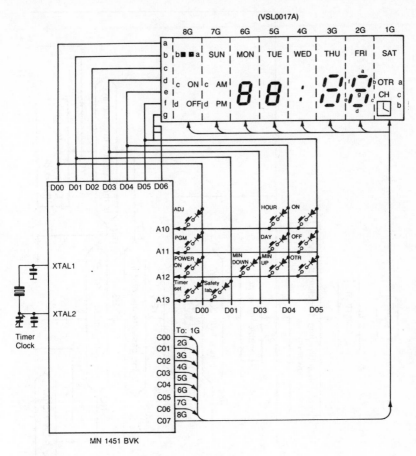

Figure 5.13 Timer key scan and display.

If a fault has been detected the following steps are performed:

1. Disable the interrupt.
2. Set up the stop mode.
3. Display the fault.
4. Call the subroutine to eject the cassette if it has been inserted.

If the power is turned off the program is returned to enable interrupt while the program sets up the stop mode if the cassette has been inserted.

The third step of the main program checks the following modes:

(a) Is it in the stop mode?
(b) Is it in REW mode?
(c) Is it in either still, PB or FF modes?

VHS system control and other circuits **161**

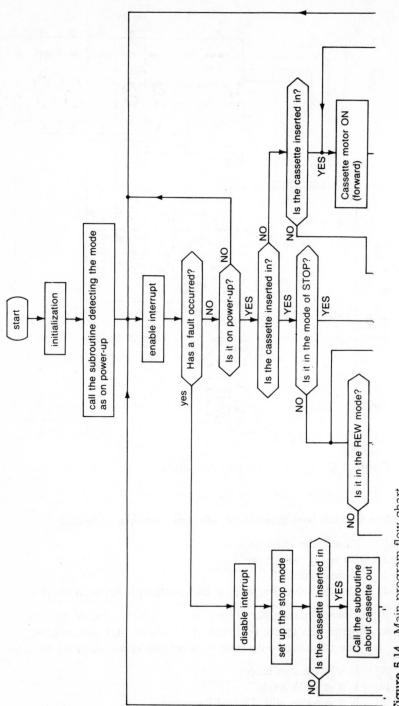

Figure 5.14 Main program flow chart.

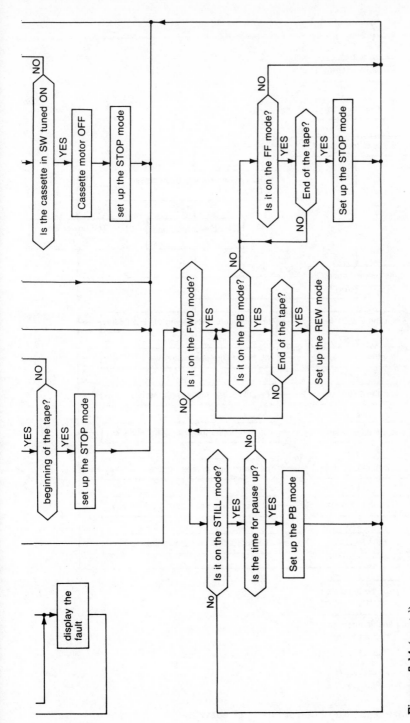

Figure 5.14 (contd)

VHS system control and other circuits **163**

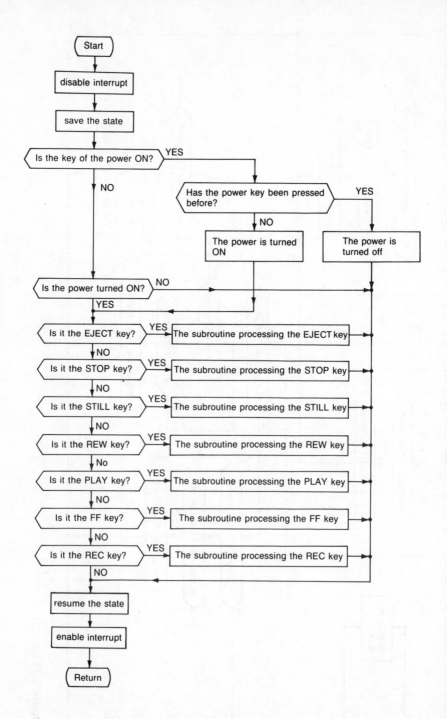

Figure 5.15 Key control interrupt service.

164 Video recorders

Table 5.3 Capability of transfer between modes

Original States	Executing states						
	EJECT	STOP	STILL	REW	PLAY	FF	REC
EJECT	0	X	X	X	X	X	X
STOP	V	0	X	V	V	V	X
STILL	V	V	0	V	V	V	V
REW	V	V	X	V	V	V	X
PLAY	V	V	V	0	0	V	V
FF	V	V	X	V	V	0	X
REC	V	V	V	V	V	V	0

Note: The symbol 0 indicates that the original mode is unchanged; V indicates that the change in mode is allowed; and X indicates that the change in mode is not allowed.

If it is in the REW mode and the tape is wound to the beginning, the stop mode is set up. If it is in the FF mode the tape is fast transported to the end of the tape and the stop mode is set up. If it is in the PB mode and the tape is wound to the end, the REW mode is set up and the tape is rewound to the beginning at the next execution of the program, and the stop mode is executed. If it is in the still mode and the pause time has elapsed, the PB mode is restored and the program branches to the PB mode at the next execution of the program, followed by setting up of the stop mode. The main program is a loop and returns to the enable interrupt to perform the next execution of the program after the stop mode has been set up.

One of the interrupt service routines checks the key control signal from the keyboard. The flow chart for this section is shown in Figure 5.15. The first step checks to see if the power-on key has been pressed after the interrupt has been disabled and the MCU register state has been saved. This check is necessary because the power is turned off and the interrupt service routine will return to the main program if the power-on key has been pressed twice.

A series of keys, EJECT, STOP, STILL, REW, PLAY, FF and REC are checked in turn after the power is turned on. If one of these keys has been pressed the corresponding subroutine will be executed and the corresponding mode is set up.

It should be pointed out that the mode has to be checked and that it is not possible to transfer between some modes as indicated in Table 5.3.

As soon as the new mode has been set up the interrupt service routine will return to the main program by resuming the state and enabling the interrupt. Another of the interrupt service routines is used for the timer. The routine sets up the initial data for the timer and turns on the timer if the following conditions have been achieved:

(a) the stop mode has been set up;
(b) the drum and capstan motors are working normally if the stop mode has not been set up;
(c) faults in the drum and capstan motors are eliminated if they occur;
(d) the reel motor is working normally, if the stop mode has not been set up; and
(e) faults in the reel motor are eliminated if they occur.

This interrupt service routine is shown in Figure 5.16.

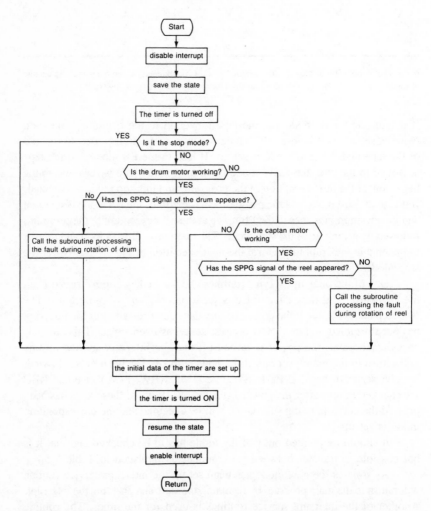

Figure 5.16 Fault checking interrupt service routine.

5.3 Other circuits in the VHS

In addition to the circuits previously mentioned, other circuits used in VHS include a tuner, a television demodulator and an RF converter. The relationship between these circuits and the video channel is shown in Figure 5.17.

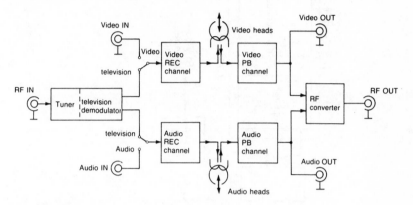

Figure 5.17 VHS RF circuit processing.

5.3.1 The tuner and television demodulator

The tuner and television demodulator are the same as those in the television receiver. The block diagram of these circuits in the V-370 is shown in Figure 5.18. It consists of a tuner, an IF filter, an IF channel IC, an IC used for demodulation in the audio channel and some filter circuits. The tuner includes an RF amplifier, a mixer and a local oscillator. The off-air signal is amplified in the RF amplifier then mixed with the local oscillator signal at the mixer and translated to the IF signal.

An electronic tuning mode is used in the tuner, i.e. it is tuned by changing the reverse bias on a varactor diode where the tuning bias is derived from a preset value stored in the memory. The receivable channels cover CH1 to CH12 in VHF and CH13 to CH69 in UHF.

The IF signal generated from the tuner goes through an IF filter, T701, to filter out the IF signal and block the noise and harmonics outside the pass band of the filter. The IF amplifier, video signal detector, video amplifier, audio signal detector and AGC for the RF and IF signals are included in IC701. The IF signal is first amplified in the IF amplifier and is then split in two directions. One path is to the detector to detect-out the video signal after it has passed through a trap circuit, T702, and an RC phase shift circuit used to remove interference from the audio carrier.

The video signal is amplified at the video amplifier and is then sent to the video REC channel via a trap, X703, for further removal of the audio

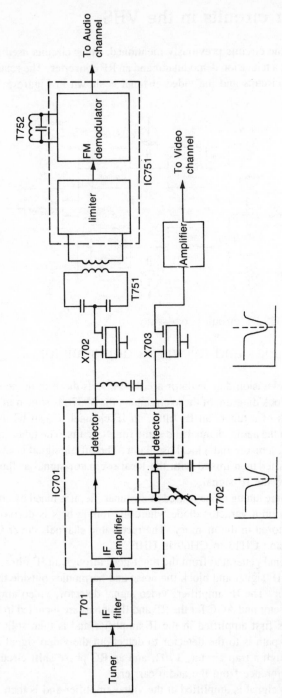

Figure 5.18 The VHS tuner and television demodulator.

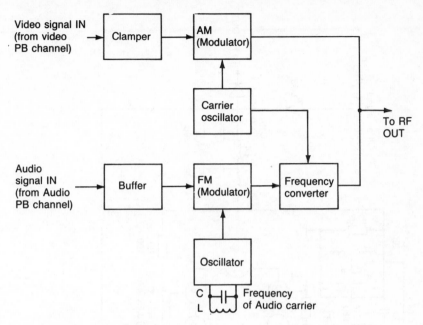

Figure 5.19 The VHS RF converter.

carrier. The other path is to a detector and filter, X702, for selecting out the audio carrier. The audio carrier is limited at a limiter and becomes the audio signal at the FM demodulator. In VHS the audio signal generated from the off-air signal is sent to the audio REC channel.

5.3.2 The RF converter

The RF converter (sometimes referred to as the RF modulator) is used to modulate the reproduced video and audio. This signal is then used to connect to a television receiver. The block diagram for the RF converter is shown in Figure 5.19. The video signal from the video PB channel is first clamped during the synchronizing period and then applied to the AM modulator to produce an amplitude modulated signal. The RF signal is generated by the carrier oscillator, oscillating at either CH3 or CH4 VHF frequency, selected by a switch (SW1 in Figure 5.20a).

The audio signal from the audio PB channel is applied to the FM modulator and a buffer to become a frequency modulated signal. The frequency of the audio oscillator is determined by a tuning circuit, which is an SAW. The SAW is shown in Figure 5.20.

The frequency modulated signal, also referred to as the second audio IF signal, is further mixed with the carrier signal from the carrier oscillator

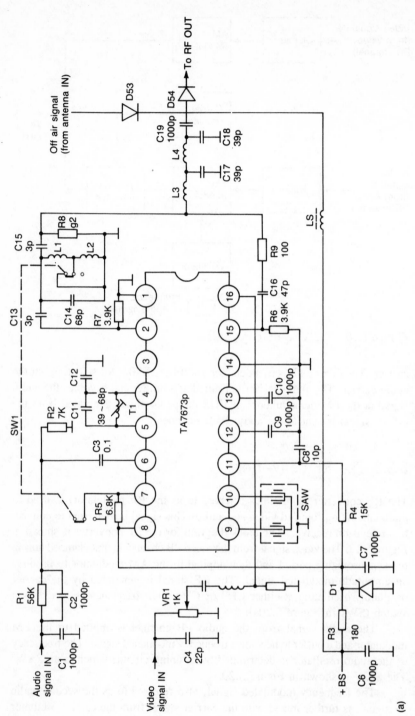

Figure 5.20 The RF converter in the VHS NV-370. (a) The circuit.

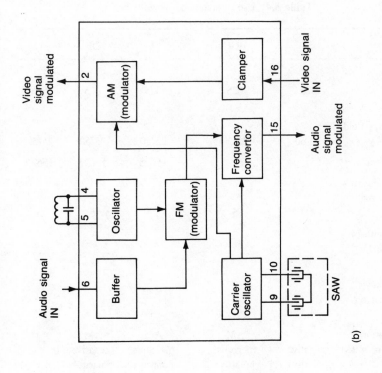

Figure 5.20 (contd) (b) The block diagram of the IC of TA7673P.

VHS system control and other circuits **171**

so that the frequency of the second IF audio is higher than that of the video carrier. The mixed audio carrier is added to the modulated video signal and is sent to the RF OUT terminal. A practical RF converter circuit, used in the NV-370EN is shown in Figure 5.20. The video and audio signals are, respectively, applied to pins 16 and 6 on the TA7673P IC, while the modulated video signal and the modulated audio signal are output, respectively, from pins 2 and 15 of the same IC. The modulated signals are added together and sent to the RF (OUT) terminal through diode D54.

When +12 V is applied to the +BS terminal on the RF converter box, the diode, D54, is turned on and the RF signal is output from the RF OUT terminal. This +12 V voltage is controlled by the television/VTR switch on the front panel.

Table 5.4 Use of standards world-wide

Characters	M	BGH	I	DK	L
Lines per frame	525	625	625	625	625
Frame frequency/field frequency	30/60	25/50	25/50	25/50	25/50
Line frequency	15750	15625	15625	15625	15625
Bandwidth of video signal (MHz)	4.2	5	5.5	6	6
Bandwidth of RF signal (MHz)	6	7	8	8	8
Frequency of audio carrier (MHz)	4.5	5.5	6	6.5	6.5
Modulation mode for video	AM (negative polarity)	AM (negative polarity)	AM (negative polarity)	AM (negative polarity)	AM (positive polarity)
Modulation mode for audio	FM	FM	FM	FM	FM
Countries	USA Japan Taiwan	Federal Republic of Germany India Malaysia Italy	UK Hong Kong	USSR China	France

As mentioned earlier, SW1 is used to switch between CH3 and CH4, so the television receiver connected to the RF(OUT) terminal of the VHS recorder should be selected to the same channel. It should be pointed out that the frequency of the RF signal output from the RF(OUT) terminal can also be one of the UHF (ultra-high frequency) channels 30 to 39 in some VHS models.

5.3.3 Television formats

It should also be noted that not only the colour television system, but that of the B/W television system should be considered when the tuner, television demodulator and RF converter are discussed. As is well known, there are three types of colour television systems used around the world:

1. NTSC (National Television Systems Committee) on subcarrier frequency 3.58 MHz. This system is used by the United States, Japan, Taiwan, Canada, for example.
2. PAL (phase alternate line) on subcarrier frequency 4.43 MHz. This system is used in the United Kingdom, Hong Kong, China, Malaysia, India, Germany, Holland, Switzerland, and other western European countries.
3. SECAM on subcarrier frequency $f_{SCA} = 4.41$ MHz, $f_{SCB} = 4.25$ MHz. Countries using this system are France, Soviet bloc countries and other European countries.

The format of the B/W television system is shown in Table 5.4. The complete designation for each of the television systems should be given as M/NTSC, D/SECAM, I/PAL, BGH/PAL, etc. With reference to Figures 5.18 and 5.20 the different points in the television demodulator and RF converter are as follows:

	T702	X703	X702	T752	SAW
D/PAL	30.5 MHz	6.5 MHz	6.5 MHz	6.5 MHz	6.5 MHz
BGH/PAL	31.5 MHz	5.5 MHz	5.5 MHz	5.5 MHz	5.5 MHz

The frequency of the channels in each of the systems is different. For example, the VHF frequencies for the tuners in D/PAL and BGH/PAL are shown in Table 5.5.

Table 5.5 VHF tuner frequencies for D/PAL and BGH/PAL

	Band I					Band II						
D/PAL	1	2	3	4	5	6	7	8	9	10	11	12
	48.5 MHz↔92 MHz					167 MHz ←——→ 223 MHz						
B/PAL	2	3	4			5	6	7	8	9	10	11	10
	47 MHz↔68 MHz					174 MHz ←——→ 230 MHz						

Chapter 6

Analogue and digital special function reproduction

With intense competition in the video recorder market over the last ten years, video recording technology has developed rapidly and has seen the advent of various kinds of new video recorders. Recorder technology has been pushed to an entirely new stage. Some new devices and techniques will be introduced in the following two chapters. These are analogue and digital special function reproduction, and standard and high band 8 mm VCR.

6.1 Introduction

The special function reproduction in VCRs is utilized in the pause or still, slow, cue and reverse operating modes, etc., and special function replay techniques are those techniques used in obtaining a noise-free picture in these modes. Reproduction may be modified by varying the replay speed and adopting some special techniques.

If it is necessary to digitize the video signal to effect the modification it is known as digital special function reproduction. If the video signal is not digitized the following problems must be considered during special function reproduction:

1. The video heads wander from the video track recorded. Figure 6.1 shows the tracks scanned by the video head in $\times 1$ normal tape speed mode (i.e. normal playback), $\times \frac{1}{2}$ speed (SLOW), $\times 0$ speed (STILL), $\times (-1)$ speed (REV) and $\times 2$ speed (CUE). It can be seen that horizontal noise band(s) occur in the replay picture because the video head track crosses the guard

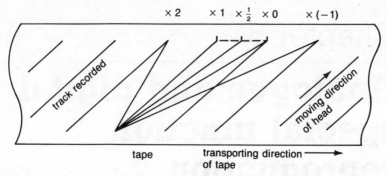

Figure 6.1 The path scanned by the video head at various tape speeds.

band(s) for the U-matic format. The same effect occurs in VHS but the video head scans across the adjacent track in this case because the guard bands do not exist in VHS. Moving the noise band(s) to the top and bottom of the screen, where they are less noticeable, or eliminating them, are the main methods of obtaining a noise-band-free picture in special function reproduction.

2. The angle of the path scanned by the video head is different from that of the recorded track at different replay speeds. A reduction in signal-to-noise ratio is hard to avoid because of azimuth loss, especially in domestic machines such as VHS, 8 mm VCR and VHS-C, etc. These machines have redesigned video heads to make use of azimuth recording for special function replay.

3. The recorded video track is the resultant of the scanning speed of the video head and the speed of the transporting tape. In special function replay the tape speed is different, so to ensure that the scanning video head path coincides with the recorded track the video head drum speed is altered. The rpm of the video head drum has to be adjusted slightly to ensure that the period of the PB H-SYNC can be pulled to that of the monitor. After this adjustment the time interval for one revolution of the drum is no longer 40 ms, thus the period of the PB H-SYNC has changed.

In addition, for recorders with two video heads which use non-segmented scanning (such as U-matic and VHS), each video head scans one track corresponding to one field. This means that it is necessary for the V-SYNC signal to be recorded on a predetermined portion of the track. The two adjacent tracks are also staggered with respect to each other, as shown in Figure 6.2.

When both heads A and B scan the same path in the still mode the PB V-SYNC pulse output from head A would lag behind that of head B and would cause flutter of the pictures generated separately by the odd and even fields. To overcome this a false V-SYNC pulse, known as the false V_D pulse, is derived in the servo system and is used to replace the PB V-SYNC pulse whose

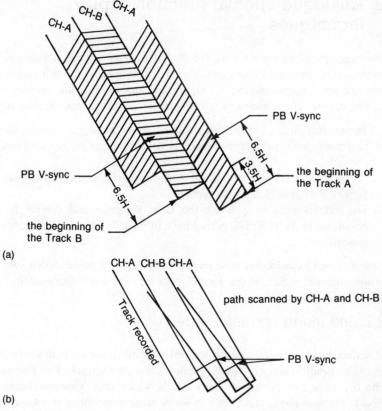

(a)

(b)

Figure 6.2 The difference in time between PB V-SYNC of heads A and B in the still mode.

period has deviated from 20 ms. The false V_D pulse should lead the PB V-SYNC pulse and should be of at least the same amplitude.

It should be pointed out that there are two prerequisites to achieving special function reproduction, as outlined above. Firstly, only recorders which use non-segmented scanning can be designed to operate in special function modes unless other measures, such as field memory techniques, are adopted. Secondly, the positions of the H-SYNC and burst on adjacent tracks have to be aligned with each other to ensure that the phase transition is continuous as the video heads cross the tracks.

These problems do not need to be considered for systems which use digital special function replay because the schemes which are adopted to achieve still picture and slow motion are totally different to those used in the analogue systems. The following paragraphs discuss separately the methods used in achieving analogue and digital special function reply.

6.2 Analogue special function replay techniques

In analogue special function replay the PB video signal is kept in analogue form throughout the total PB channel. The fast, slow motion and still picture modes are achieved by varying the tape speed and result in the problems described earlier. Three measures are adopted to overcome these problems.

1. The control circuit for the still mode has to be arranged to make the tape move slowly to the appropriate position after the pause key has been pressed.
2. The new video head(s) for the still mode have to be redesigned to meet the needs of special function replay.
3. The circuits used to generate the false V_D pulse and correct for deviations in the H-SYNC period have to be added to the drum servo system.

The method used for noise-free slow motion reproduction is based on that used for a noise-free still picture, so that for still mode reproduction is discussed first.

6.2.1 Still mode optimum video scan

As described earlier, when the tape is stopped in the still mode the path scanned by the video head(s) cannot completely follow the recorded track. This results in the occurrence of horizontal noise bands which spoil viewing during playback. For U-matic format VCRs, in which azimuth recording is not used, the video signal can be reproduced even though the video head path suffers deviation and scans the adjacent track. A horizontal noise band appears in the still picture as a result of the video heads scanning the guard band between tracks. The path scanned by the video head is random for random still images if no measures are taken to control the guard band noise. Figure 6.3 shows two typical positions of the scanned path.

Figure 6.3b shows the path scanned by the video head which intersects the recorded track in the mid-part only. The top and bottom of the path lies in the guard band so that noise bands would appear at the top and bottom of the screen. This results in minimum interference on the PB picture.

Figure 6.3a shows the case where the path scanned by the head intersects the guard bands at their mid-point causing the noise bands to appear in the middle of the still picture. This is clearly the situation which needs to be avoided.

For domestic VCRs, such as VHS, Betamax, etc., the interference from guard bands does not exist. However, since azimuth recording is used deviation of the video head from the recorded track results in scanning of the adjacent track. The noise bands in the still picture are therefore a result of azimuth

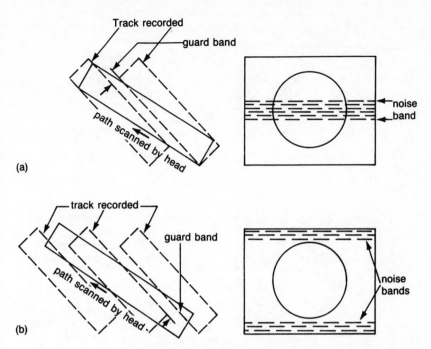

Figure 6.3 The relationship between the guard band and the path scanned by the video heads of VCR with U-matic format in the still mode: (a) the noise band appears in the middle of the screen, (b) the noise band appears at the top and bottom of the screen.

loss as the head is scanning the adjacent track. Similarly, the noise bands produced in this case are placed at random. If the capstan servo contains still control, the tape is moved slowly to stop in a position in which the noise bands cause minimum picture interference. These are shown in Figure 6.4a, b or c, and are the optimum positions after the PAUSE key is pressed.

In Figure 6.4a the two heads scan the same track, for example track B, and the PB signal is output from one video head only, head B. The still picture is thus provided by one video track, or one field, and is known as the field-still mode.

Figures 6.4b and c show the case where each video head scans two tracks. The still picture is produced from the beginning of track A and the end of track b, and is known as the frame-still mode.

6.2.2 The still control circuit

The still (or pause) control circuit is fundamental in achieving a noise-free still and slow motion picture and is discussed first. The pause control circuit

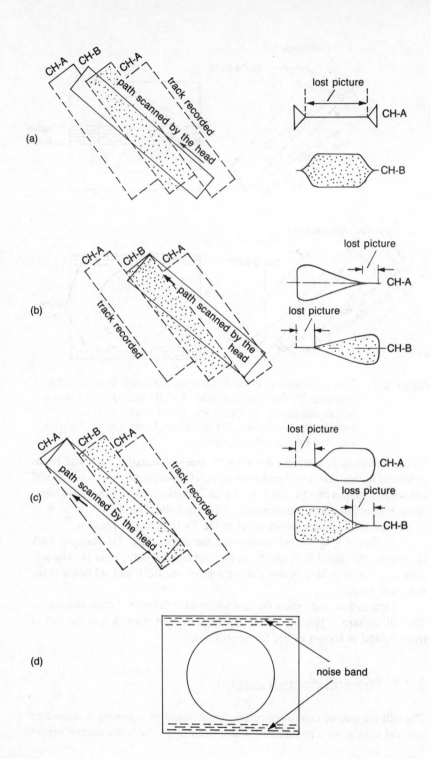

in a U-matic VCR has been discussed in detail in Chapter 4 so the discussion here is concerned with that in the domestic VCR.

One scheme uses a 'window' in which the noise band is permitted to appear on the screen. After the still mode is selected, the position of the noise band is detected and the capstan motor is driven forward by a position pulse, derived from the still control circuit, until the noise band is positioned in the window. The still control circuit has been produced in an integrated circuit, such as the BA841 and BA855, etc., and also as part of the large scale integration of the servo system.

Figure 6.5 shows the still control circuit in the VT-8000, which also consists of generation and suppression circuits for the shift noise pulse.

When the still mode is selected, a still(H) signal, originating in the control system, is inverted to a low logic level through IC4. This signal is fed to IC6-1 to unlock the shift noise pulse generator. It is also fed to IC6-13, which is an R−S trigger consisting of two NOR gates. The noise band detected by the drop-out detector is shaped and amplified by Q4−Q7 and IC4 and is inverted to form a negative pulse at Q10. This is used to indicate the noise position on the screen.

When the noise band does not fall into the window set by IC7 and Q3, i.e., the two pulses at IC6-8 and IC6-9 are not coincident in time, the output of IC6-10 is at a low logic level. The R−S trigger is therefore not activated and Q4 remains turned off. The shift noise pulse from IC6-3 is thus not suppressed and drives the capstan motor until the pulse indicates that the noise band falls within the window.

When the noise band is moved into the window and negative pulses are received at IC6-8 and IC6-9 at the same time, IC6-10 outputs a positive pulse to trigger the S-terminal of the R−S trigger. A high level is then output from IC6-11 which turns Q4 on. The shift noise pulse is then by-passed causing the capstan motor to stop in the predetermined position.

The shift noise pulse and the predetermined window are both generated from the head switch (H-SW) pulse. The shift noise pulse is generated only when the still mode is selected, i.e. when the still(H) signal appears. Each frame generates one shift noise pulse. Its width is adjusted by R3 and its

Figure 6.4   The relationship between the guard band and the path scanned by the video heads of VHS in the still mode: (a) no output from the CH-A head; (b) no output from the CH-A head at the bottom of the screen and no output from the CH-B head at the top of the screen; (c) no output from the CH-A head at the top of the screen and no output from the CH-B head at the bottom of the screen; (d) the noise bands appear at the bottom and top of the screen in figures (a), (b) and (c).

Analogue and digital special function reproduction   **181**

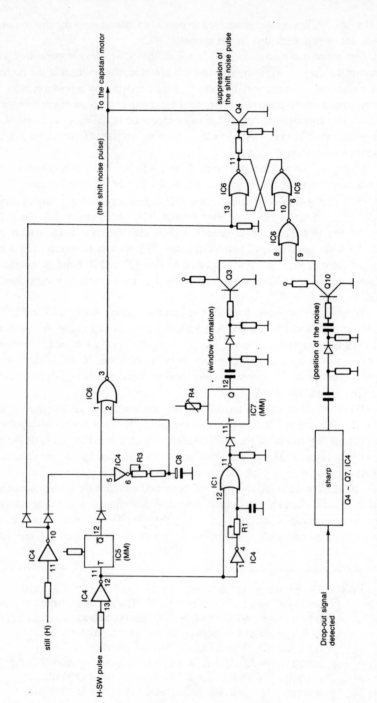

Figure 6.5 The still control circuit in VT-8000.

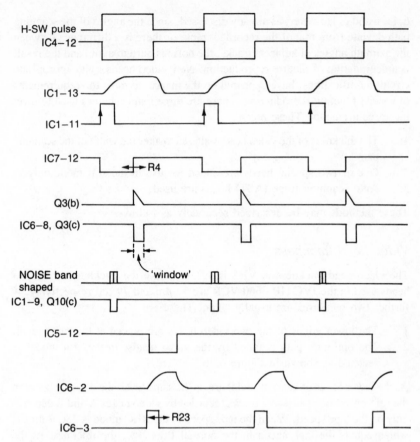

H-SW pulse
IC4–12

IC1–13

IC1–11

IC7–12

Q3(b)

IC6–8, Q3(c)

NOISE band
shaped
IC1–9, Q10(c)

IC5–12

IC6–2

IC6–3

R4

'window'

R23

Figure 6.6 The waveforms of Figure 6.5.

waveform is shown in Figure 6.6 at a point when the still mode is just set up. The capstan motor does not stop and rotates a bit, driven by the shift noise pulse at each frame, until the noise band moves step by step to the bottom of the screen, out of view.

At the output of the NOR gate, IC1-11, the inverted H-SW pulse becomes a positive pulse which is delayed by a monostable, IC7. This produces a window pulse through Q3 and the position of the window can be adjusted by R4. It can be seen from Figure 6.6 that in order to move the noise band to the bottom of the screen the window pulse has to be adjusted to a position as near as possible to that prior to the head switching point.

6.2.3 Video heads for special function playback

The optimum video heads scanning path can be selected as shown in Figures

6.4a, b and c. However, as already discussed, since the angles of the scanning path deviate from that of the recorded track, or there is a difference between the azimuth angles of adjacent tracks, the noise is hard to avoid and the result is a degradation of picture reproduction. New video heads, with appropriate circuit modifications, have appeared on the market to meet the requirements of special function reproduction. There are three main schemes used to meet these requirements. These are:

1. The thickness of the video head is altered to alter the width of the scanning path.
2. One or two special heads are added for use in the still mode only.
3. Auto scanning track (AST) heads are used.

These methods may be described separately as follows.

Video head thickness

The video head thickness in VHS is 49 μm, while the thickness of the two heads used in the JVC HR-3660 VCR are 59 μm and 79 μm respectively. A further two measures are usually taken. These are:

1. The lower edges of the two video heads are placed at the same level.
2. The optimum path scanned by the video heads, in the still mode, is selected as shown in Figure 6.7b.

As the lower edges of heads A and B are on the same level, the distance between the reference line of adjacent tracks recorded by video heads A and B depends only on the tape speed. When the tape speed in the REC mode is 23.39 mm/s, which equals the tape speed in the normal PB mode, the thickness of the recorded track is still 49 μm. In fact, during recording, the track width recorded by head A is 59 μm but the successive track recorded by head B overlaps it by 10 μm so only 49 μm remain. Similarly, the 79 μm width track recorded by head B is overlapped by head A by 30 μm resulting in a track width of 49 μm also. Thus, as the tape travels at normal speed, the recorded track width is 49 μm in spite of the fact that the video heads of the HR-3660 VCR are 59 μm and 79 μm thick.

During reproduction, the tape stops in a position such that the two video heads just sit over recorded track A, as shown in Figure 6.7b. The width of the path along which head A picks up the signal is 59 μm, which is wide enough to cover the top and bottom areas of track A, even if, during the still mode, the angle of the scan path deviates slightly from that of the recorded track. This results in an improvement in the quality of the PB A field signal.

Head B cannot pick up a signal from track A since the reference lines of heads A and B coincide and the scan path of head B is nearly twice as wide as one of the recorded track, i.e. the scan path of head B covers most of the area of the next recorded track B, so the field signal is also reproduced normally.

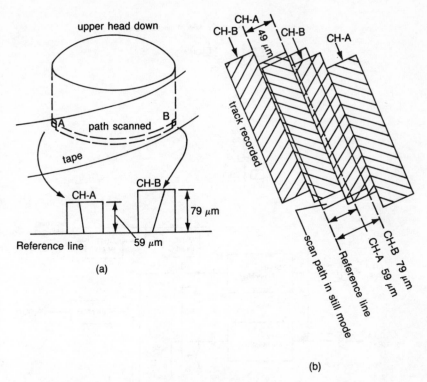

Figure 6.7 The thickness of the video head is altered to alter the width of the scanning path.

With this scheme, the problems of PB still picture degradation, resulting from the difference between the angle of the recorded track and that of the scanning path, and the problems resulting from the fact that both heads cannot pick up the PB signal from one track because of azimuth angle differences, can be solved at the same time.

Use of special heads

Another scheme used to solve these problems is to add one or two special heads for reproduction in the still mode.

Hitachi VT-136E and VT-426E VCRs are examples of machines which provide one extra head for still mode reproduction. This is shown in Figure 6.8. In addition to CH-A and CH-B video heads mounted on the upper drum for normal PB and REC modes, a special function replay head, whose gap azimuth is the same as that of head B, is attached to the drum near head A for use in the still and slow motion modes.

In the still mode the tape is controlled to halt in the exact position to ensure that the two video heads can alternately pick up the signal from track

Analogue and digital special function reproduction   **185**

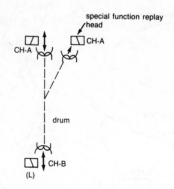

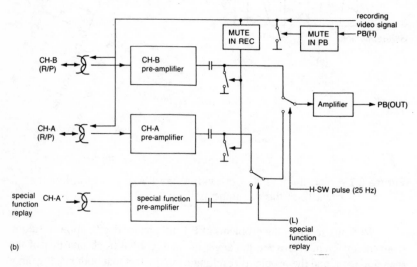

Figure 6.8 A special function still replay head (one extra head).

B. Therefore, the special replay head, CH-A′, works with head B in place of the CH-A head. As the azimuth of head A′ coincides with that of head B the signal can be output in either the B or A field periods.

Using a special replay head can also result in an improvement in the S/N ratio in the still mode. The vertical resolution is not improved because the PB signal in the two field periods are picked up from the same recorded track, i.e. track B.

The special replay head, CH-A′, can be mounted with head A on a head base to form a 'double gap' head. This type of head assembly is found in the Sony SL-HR60 VCR.

The V-8600 VCR, an early product by Toshiba, can be used to illustrate the addition of two special heads. This situation is shown in Figure 6.9. On the upper drum, CH-A and CH-B are standard heads used for normal REC

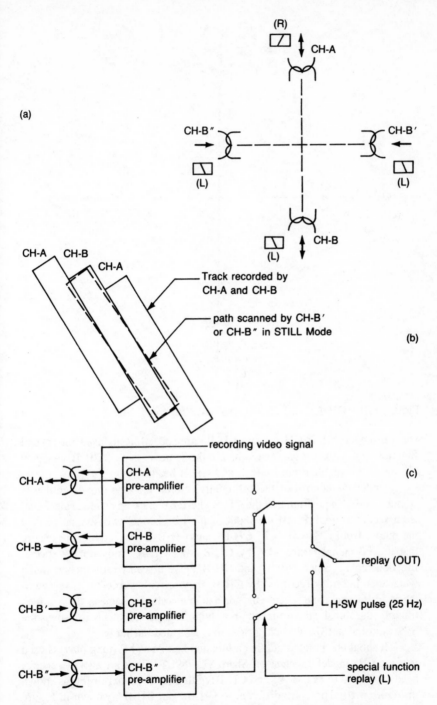

Figure 6.9 Special function replay using two additional heads.

Analogue and digital special function reproduction **187**

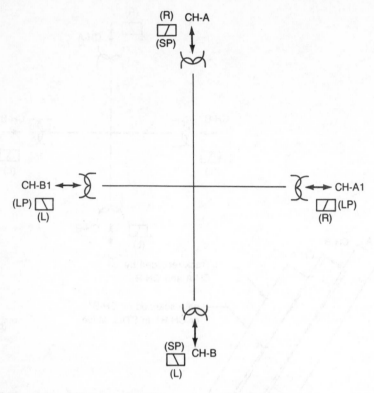

Figure 6.10 CH-A1 and CH-B1 are used for the LP mode.

and PB modes, while CH-A' and CH-B' are special heads used for special function replay. Their gap azimuths are the same as that of CH-B, but they are much thicker than both CH-A and CH-B heads.

Under the control of the still control circuit the tape moves until it is stopped in such a position that the CH-B' head can pick up an alternate signal from recorded track B. At the input of the PB channel two switches select the output from CH-A, CH-B, CH-B' and CH-B", respectively, to ensure that the PB signal is provided by CH-B' and CH-B" only during special function replay. Since CH-B' and CH-B" give a scan path which is much wider than the track recorded by CH-B, they provide sufficient coverage of the recorded track, and because the gap azimuth is also the same as the recorded track B, the signal can be picked up in both field periods. This improves the S/N ratio but not the vertical resolution, as stated earlier.

It should be pointed out that a function known as LP (long play) is used in some VHS models, such as the Sharp VC-789ET. There are also two special heads, referred to as CH-A1 and CH-B1, fitted on the upper drum, as shown in Figure 6.10. These are different to CH-B' and CH-B", shown in Figure 6.9a. They are different in three aspects:

1.   The azimuth angles of CH-A1 and CH-B1 have opposite senses of rotation in relation to the perpendicular of the video track, while both CH-B′ and CH-B″ have the same azimuth angle as CH-B.
2.   CH-A1 and CH-B1 are thinner than CH-A and CH-B, while CH-B′ and CH-B″ are thicker.
3.   CH-A1 and CH-B1 can be used for recording and replay while CH-B′ and CH-B″ heads are used for replay only.

The LP function is related to the SP (standard play) function. The tape is played at standard speeds in SP mode, while the tape speed is reduced and the recording density is increased resulting in a prolonged playing time in LP mode. The playing time is usually doubled in the LP mode so a standard three-hour tape can record or play back for six hours in the LP mode.

There are five heads fixed on the drum in the Panasonic NV-788 as shown in Figure 6.11. Two of the heads, CH-A and CH-B, are the same thickness (70 μm), but the azimuths have opposite rotation in respect to the tape perpendicular. These heads are used for REC and PB in the SP mode.

Two other heads, CH-A1 and CH-B1, have different thickness (30 μm and 40 μm, respectively) and have opposite azimuth angles. They are used for REC and PB in the LP mode. The final head, CH-B′, has a thickness of 70 μm and the same azimuth as CH-B and is used with CH-B for special function replay.

In the SP mode, the tape speed is 23.39 mm/s and the width of the recorded track is 49 μm, though the thickness of CH-A and CH-B is 70 μm. This can be seen from Figure 6.11b and is analogous to the situation shown in Figure 6.7.

In the LP mode, the tape speed is half that in the SP mode, i.e. 11.695 mm/s, and the track width is also half that in the SP mode, i.e. 24.5 μm, even though the thicknesses of CH-A and CH-B are 30 μm and 40 μm respectively. The replay time can therefore be doubled.

In this recorder there is a control circuit for still and slow motion replay. In SP still mode the tape is controlled to stop in the position shown in Figure 6.4a. CH-B and CH-B′ alternately pick up the signal from track B. This is known as a field-still scheme.

There is also a noise-free still function in the LP mode. The tape is controlled to stop in the position shown in Figure 6.4c. In this case either CH-A1 or CH-B1 picks up the signal from the recorded video track. Since the tracks scanned by CH-A1 and CH-B1 are much wider than the tracks recorded, the S/N ratio of the PB signal is satisfactory. This is a frame-still scheme.

Auto scanning track heads

The auto scanning track (AST) head assembly consists of a video head attached to a piezoelectric carrier which can move the head up or down according to

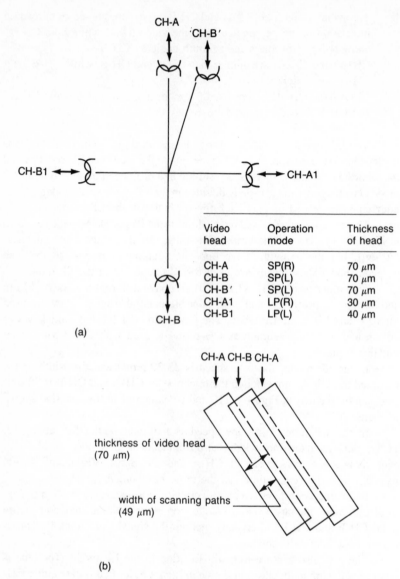

Figure 6.11 The video heads and tracks in Panasonic NV-788.

a voltage applied across the crystal. This is shown in Figure 6.12.

In order that the head can move vertically, the carrier is made of two piezoelectric crystals (the upper and lower plates). A grounded soft metal plate is sandwiched in between the plates. When a voltage is applied across the carrier, voltages of opposite polarity appear across each of the crystal plates. If the voltage across the carrier is positive at the top and negative at the bottom,

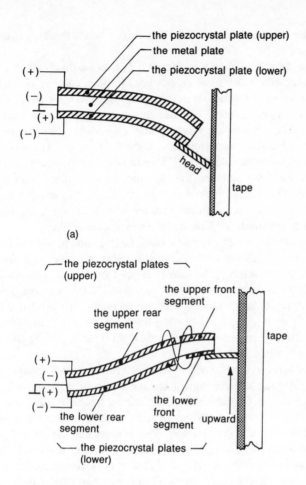

(a)

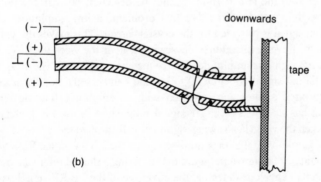

(b)

Figure 6.12 The configuration of the AST head assembly.

the carrier will bend downwards. The carrier bends upwards if the polarity is reversed.

The device, as shown in Figure 6.12a, has a disadvantage. When the carrier bends the video head cannot keep in perpendicular contact with the tape, thus causing a reduction in PB signal level. To solve the problem, each of the two crystal plates is divided into two segments, the front and rear segments. The segments are wired together crossed, i.e. the upper rear segment is connected to the lower front one, and the lower rear is connected to the upper front. Since the voltages across the front and rear segments are of opposite polarities the front part and the rear part of the carrier bend in opposite directions. The whole carrier forms an 'S' shape. This ensures that the tape maintains perpendicular contact with the head when the head moves either upwards or downwards. This is shown in Figure 6.12b.

In order to realize automatic track finding, and hence a noise-free still picture, a servo system is also needed to work with the AST heads. When the still mode is selected the tape is stopped by the microcomputer control system in the position as shown in Figure 6.4b. In this position, the scan path of head A coincides quite well with the beginning of recorded track A, but deviates from it to the left along the track length. The scan path of head B deviates from the recorded track B to the right at the beginning and gets closer to the track along its length.

The servo system provides each head carrier with a voltage of opposite polarity, forming a correcting voltage across carrier A from low to high, and from high to low for carrier B. Thus, during scanning under the control of the ATF servo system, head A moves upwards to scan track A and head B moves downwards to scan track B. This ensures that the PB still picture has the same S/N ratio and vertical resolution as that of a normal PB picture.

Making use of the fact that deviation of the scanning path from the video track reduces the PB signal, the AFT servo system detects the amplitude of the PB RF signal and provides a correcting voltage to force the video head to follow the recorded track. However, with this scheme the deviation direction of the head, to the left or right, cannot be detected, nor can the polarity of the correcting voltage. To solve this problem, a low amplitude reference oscillating signal is provided for the crystal carrier. This makes the video head swing in the traverse direction while scanning along the video track. This means that the PB signal is amplitude and frequency modulated. The frequency of the oscillating signal is below 200 kHz and can be separated from the PB C signal through a filter and removed with an amplitude limiting circuit. It therefore has no effect on the Y signal modulated in the PB Y channel.

When the video head swings symmetrically around the centre of the track, the head is following the track quite well, and the PB RF signal from the video head would be that shown in Figure 6.13a. Using a shaped reference oscillating signal and an envelope detector, the envelope of the PB RF signal waveform is detected. The positive and negative voltages of the sampled envelope are

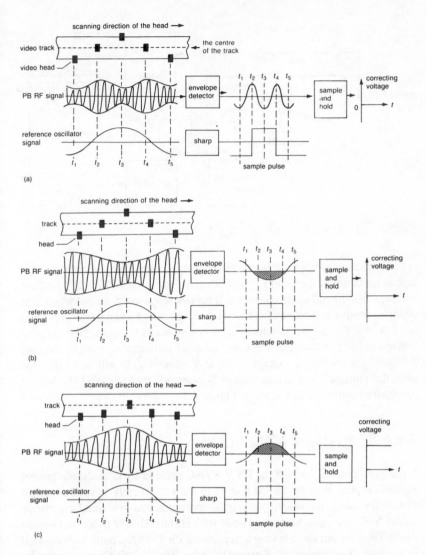

Figure 6.13 Detection of the deviation direction of the video head.

equal to each other and the correcting voltage provided by a holding capacitor is zero.

Figure 6.13b and c show the case where the video head deviates from the video track to the left and right respectively. In Figure 6.13b, the voltage detected from the sampled waveform is negative and it is used to drive the video head to the right. Similarly, the voltage detected from the sampled waveform is positive and is used to drive the video head to the left. The combination of these functions produces automatic track finding using circuitry as shown in Figure 6.14.

Analogue and digital special function reproduction   **193**

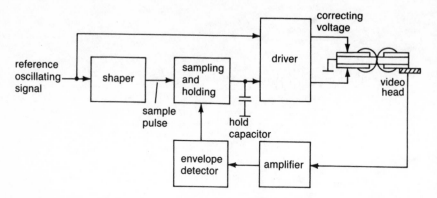

Figure 6.14 The circuit driving the AST head.

The tape control used by the microcomputer control system to stop the tape in the position shown in Figure 6.4b can now be examined. As described earlier, when the still mode is selected, if the tape stops just in the predetermined position, head A will require the least correcting voltage when it starts scanning track A. So, the microcomputer control system needs to detect the correcting voltage at the beginning of the scan and send instructions to drive the tape a little if the correcting voltage is not at a minimum. It will drive the tape until the minimum correcting voltage is detected, thus stopping the tape in the desired position, as shown in Figure 6.4b.

6.2.4 The false V-SYNC pulse

It has been pointed out that the PB V-SYNC pulse is variable during special function replay, causing the PB picture to flutter vertically to such an extent that the monitor is not able to pull it into synchronization. It is usual to generate a false V-SYNC pulse using the H-SW (in VHS) or RF-SW pulse (in U-matic format) in the drum servo system to replace the PB V-SYNC pulse whose period is unstable. The method used to generate the false V-SYNC pulse in the U-matic format was discussed in Chapter 4 and the scheme used in VHS format machines follows.

Figure 6.15 shows the generator used to produce the false V-SYNC pulse in the NV-G30EN VCR. (Note that in the original machine the words 'artificial V-SYNC' are used instead of 'false V-SYNC'.)

Ignoring the delay/direct circuit for the moment, the H-SW pulse is divided into two monostables, MM1 and MM2, to produce one negative pulse per field from the NAND gate. Monostable MM3 is used to set the width of the false V-SYNC pulse.

The delay/direct circuit is marked on the original diagram as SPECIAL

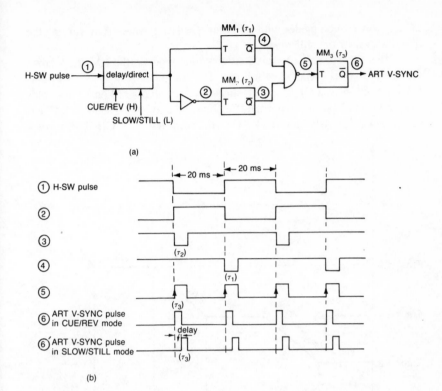

Figure 6.15 Generation of the artificial V-SYNC pulse.

PB MODE INPUT SELECT, which means it is used to select the input method for the H-SW pulse according to the special PB mode. It is used to delay the H-SW pulse for still and slow motion modes, while letting the H-SW pulse pass directly through for forward and reverse search modes.

In special function replay, besides providing the false V-SYNC pulse, the drum servo needs to adjust the rotational speed of the drum slightly in order to minimize the variation in the PB H-SYNC period. In the VO-585OP a PB synchronizing pulse has replaced the phase servo circuit to drive the drum motor. In the NV-G30EN an H-SWAY pulse formed from the H-SW has been added to the drum phase servo. Whether it is the PB SYNC or the H-SW SYNC pulse which varies with the rotating speed of the drum, all the above adjustments are based on the feedback principle.

6.2.5 Noise-free slow motion mode

Noise-free slow motion mode is based on the noise-free still mode. It is achieved by repeatedly switching between the noise-free still mode and the normal PB mode. Slow motion modes with different speeds are achieved by varying the time interval of the still mode.

The $\frac{1}{4} \times$ normal tape speed slow motion mode in the NV-788, shown in Figure 6.16, is executed as follows. The still mode is held for two fields (2V), i.e. CH-B and CH-B$'$, each successively scan track B once. Six fields (6V) are then played in the normal PB mode in the following sequence: CH-A scans track A1, CH-B scans track B1; CH-A scans track A2, CH-B scans track

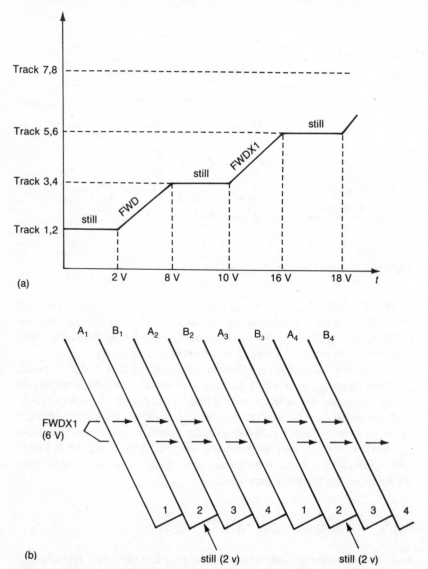

Figure 6.16 Slow ($\times \frac{1}{4}$) mode in NV-788.

B1; CH-A scans track A2 and CH-B scans track B2. This means that the tape will spend the time for six fields in crossing only two tracks. The total time, including the two still fields, is eight field time periods.

$$\frac{2 \text{ tracks transported}}{\text{time for eight fields}} = \frac{1}{4} \times \text{normal tape speed}$$

Generally speaking, the tape speed in the slow motion mode depends on the time interval for the still picture. For example, $\frac{1}{30} \times$ normal tape speed should be:

$$\frac{2 \text{ tracks transported}}{54 \text{ fields for still mode} + 6 \text{ fields for normal PB mode}} = \frac{1}{30}$$

In order to achieve noise-free slow motion, the requirements are as follows:

1. A stepping motor used as the capstan motor should be of low rotary inertia so it can be stopped and started easily.
2. Since the capstan motor needs a high starting current, the driving pulse should have a staircase waveform as shown in Figure 6.17.

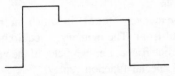

Figure 6.17 The capstan stepping pulse for the slow motion mode.

3. The still control circuit is usually implemented using a large scale integration (LSI) chip such as the BA841 and the BA855, or using a microprocessor, such as the UPD-1511AC020, which is a four-bit NMOS MUS (monolithic), to incorporate the still/slow motion control.
4. The drum servo circuit is required to provide the false V-SYNC pulse, and also make fine adjustment of the drum's rotation speed to ensure that the PB H-SYNC pulse can be pulled into synchronization by the monitor.

6.3 Digital special function replay

The main characteristic of digital special function reproduction is that the video signals are digitized in the PB channel, and the still, or other special function modes, are achieved by changing the time ratio between the signal being written into and read out of memory. From the storing and reading of data viewpoint, the analogue and digital replay schemes are based on the same principle. That is, storing the video signals at the normal rate and reading them out at a variable rate. In the case of storing and picking up analogue video signals, using the tape as the storage medium, changing the tape speed changes the rate of picking up information. In the case of digital video signals, using semiconductor memories as the storage medium, changing the speed of reading out information also changes the rate of picking up information.

The Toshiba DV series VCR, including the DV-80, DV-90D, DV-93D, DV-98C, can be used as examples to illustrate the operating principles of digital special function replay.

6.3.1 Digital special function replay scheme

In the digital special function replay scheme, the PB video signals are digitized and stored in semiconductor memory. Various digital special effect pictures can be obtained by varying the speed of picking up the data. The scheme, in principle, must involve the following four parts:

1. An analogue to digital converter (A/D C) which converts analogue signals into digital signals.
2. A semiconductor memory (RAM) in which the digital signals can be stored or read out from. The memory is required to be able to at least store one field or one frame, i.e. two fields of the video signal are needed to realize digital special function replay.
3. A digital to analogue converter (D/A C), which converts the signals read out from the memory into analogue signals.
4. A memory controller and address generator, which controls the A/D C, D/A C and memory so that they can be operated in a predetermined order, and the address of the memory storing the field of video signal can be allocated.

Figure 6.18 shows the block diagram of the digital special function replay scheme in the Toshiba DV series VCRs.

The switch, ICF02, is selected to the normal terminal by a control signal, $D/\bar{A}$, which is a low logic level. This is output from the microcomputer consys in the normal PB mode. The PB signal is output directly through switch ICF02.

When the still mode is selected, the level of the control signal is changed from a low logic level to a logic high and the switch transfers to the PAUSE terminal. The PB video signal is fed through the digital still circuit before it is output via the switch, ICF02.

The signals input to the memory controller, shown in Figure 6.18, are as follows. The system clear signal (SYS CLR) is a positive tip pulse and is used to reset the timer, counter and dividers in the memory controller. When the still mode is selected they are reset first before operation starts.

The read/write control signal (RE/$\overline{WE}$) remains at a high logic level

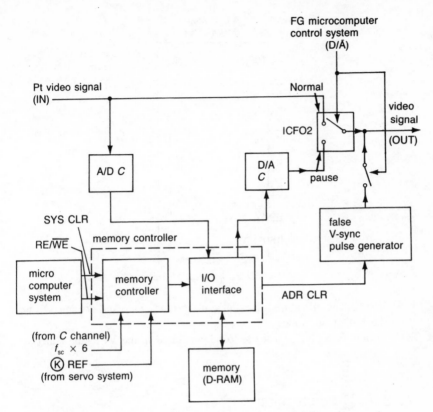

Figure 6.18 The block diagram of a digital special function replay.

throughout the still mode. Switching from low to high level is timed by the K REF signal. The K REF signal is the frame reference ($\frac{1}{2} V_D$) signal mentioned previously. It is a symmetrical square wave with a frequency of 25 Hz. It is used here to form the false V-SYNC pulse and to time the switching position of the RE/WE mode. The video signals are always written in or read out one field at a time and the RE/$\overline{WE}$ switching positions are always timed to occur at the rising edge of the K REF signal, which is near the V-SYNC pulse. The SYS CLR and RE/WE signals originate in the microcomputer control system while the K REF signal is derived from the servo system.

After being multiplied by a factor of six, the subcarrier (f_{SC}) from the chroma channel is sent to the memory controller where it is split and sent to the A/D C, the D/A C and the I/O interface to be used as a clock pulse. It is also used to generate the address for the memory. Since the memory address is reset near the V-SYNC pulse, and the memory capacity is not large enough to store a complete field of the video signal, some of the signal near the V-SYNC pulse is not stored. (The memory is large enough to store only 308 lines.) To make up the deficiency, two measures have to be taken:

Analogue and digital special function reproduction **199**

A/D *C* chip
(MB 40576)

9	V_{CCD}	GND	8
10	V_{CCA}	CLK	7
11	V_{RT}	D_1	6
12	V_{IN}	D_2	5
13	V_{RB}	D_3	4
14	V_{CCA}	D_4	3
15	V_{CCD}	D_5	2
16	GND	D_6	1

Figure 6.19 A/D *C* chip used in the digital still circuit.

1. Four of the 308 stored lines have to be read twice and arranged to be the last of one field.
2. A false V-SYNC signal has to be generated in the VCR and added to the output.

6.3.2 A/D conversion

Converting the analogue video signal into a digital signal usually consists of sampling, quantizing and decoding. Integrated circuits are available to perform A/D conversion. Some of these include all the three steps needed for A/D conversion while some only include the last two steps and sampling has to be performed by additional circuits. The A/D conversion chip used in the Toshiba DV series VCR includes all three steps and its pin configuration is shown in Figure 6.19.

After the video signal is input via the V_{in} terminal (pin 12) it is sampled at the clock pulse rate input at the CLK terminal (pin 7), resulting in a series of narrow pulses whose period is the same as that of the clock pulses. The amplitude of the sample pulses varies with, and is equal to, the amplitude of the video signal being sampled. The sampling frequency is usually triple that of the subcarrier frequency ($3f_{SC}$) so that the signals, after D/A reconstruction, do not suffer distortion and beat frequency interference.

Figure 6.20a shows a series of narrow pulses which is output from a sample hold circuit after an analogue signal has been sampled. Each pulse is known as a sample point pulse. Generally, in A/D conversion, the number of levels (N) into which an analogue signal is to be quantized, and the number of data bits (n) corresponding to the output data have the following relationship:

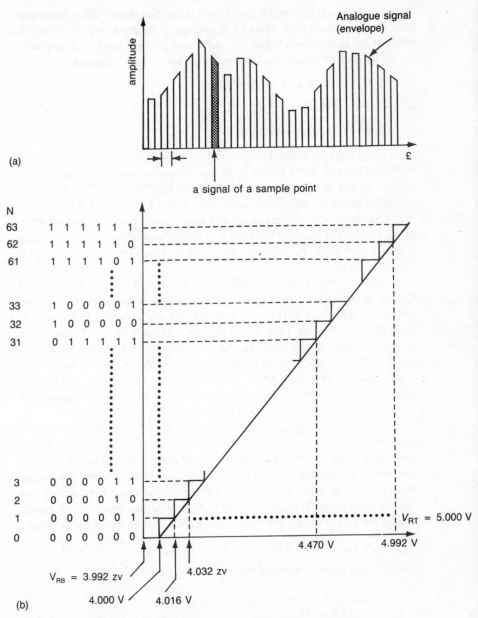

(a)

a signal of a sample point

(b)

V_{RB} = 3.992 zv

4.000 V   4.016 V

4.032 zv

V_{RT} = 5.000 V

4.470 V   4.992 V

Figure 6.20 Illustration for sample quantization and encoding.

$$N = 2^n$$

The envelope of the sample point pulse is smooth while the waveform reconstructed from the quantized signal is stepped, with the difference between them being known as quantization noise. Obviously, the more levels, or bits, there are the less is the magnitude of the quantization noise. The relationship between the peak-to-peak value of the sample point pulse and the effective value of the quantization noise, i.e. the signal-to-noise ratio, is determined by the number of quantizing bits (n) in the following expression:

$$S/N = 6.02 \times n + 10.8 \text{ (dB)}$$

which can be approximated to:

$$S/N = 6 \times n + 10.8 \text{ (dB)}$$

For ordinary digital video devices, such as digital time-base correctors and digital VTRs, etc., eight bits ($n = 8$) is sufficient to ensure adequate picture quality. In this case, the number of quantized levels is $N = 2^8 = 256$, with the signal-to-quantization noise ratio being $6 \times 8 + 10.8 = 58.8$ (dB).

For digital still pictures the quality requirements can be relaxed slightly. In the Toshiba DV series of VCR the A/D C chip used in the digital still picture circuit is the MB40576, which has 64 quantizing levels and can encode six-bit data (D_1 to D_6).

The PB video signal is usually about 1 volt peak-to-peak and the A/D C chip requires a sync-tip level of 3.992 V, so the signal has to be clamped before it is sent to the A/D C chip to ensure that the level of the input signal, from the sync-tip to the peak white, is from 3.992 to 4.992 V. Figure 6.20b shows the 64 quantizing levels and the corresponding six-bit encoded data in the A/D C chip, where 000000 corresponds to 3.992 V, and 000001 is the binary encoded data for 4.000 V. Each increase of 16 mV adds 1 to the six-bit binary count, so 16 mV is the quantizing step. The highest quantizing level is 4.992 V corresponding to 111111.

Thus, before being sent to the A/D C chip, the PB signal has to go through a clamper to restore the d.c. level. In the chip, it is sampled by a sample and hold circuit and becomes a series of sample point pulses. These pulses are quantized and encoded into equivalent six-bit binary data groups. The output from the chip is six-bit parallel data, each group of which corresponds to a sample point pulse. The time interval between two successive output groups equals the period of the sampling clock pulse. The sampling frequency is:

$$3 \times f_{SC} = 13.3 \text{ (MHz)}$$

and the time interval between two successive output groups is:

$$T = \frac{1}{13.3 \times 10^6} = 75 \text{ ns}$$

Figure 6.21 shows the input and output interface circuits for the A/D C chip. After the PB signal (1 V peak-to-peak) is output from the buffer (QV21), its d.c. level is decoupled by capacitor CV05 before entering the clamper, consisting of V23 and V01. The emitter voltage (V_B) of the buffer (QV23) is dependent on the bias resistors, RV08, RV07 and QV22 (used for

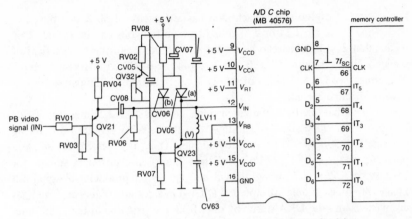

Figure 6.21 The input and output circuits around A/D C chip.

temperature compensation), and is required to be 3.992 volts. When $V_B =$ 3.992 volts, diode DV01(a) is turned on and diode DV01(b) turns on only when the voltage reaches that of the sync-tip pulse. This means that the sync-tip pulse is clamped to $V_B = 3.992$ volts. V_B is also sent to the VRB terminal of the A/D C chip where the sync-tip pulse is equal to the lower limit of the quantization required in the chip. The upper limit terminal of the A/D C chip is connected to $+5$ V, corresponding to the highest quantization level, 4.992 V. The clamped video signal is input to the chip via the V_{in} terminal (pin 12). The sampling clock pulses ($3 \times f_{SC}$) are sent from the ACLK terminal (pin 66) of the memory controller and are connected to terminal CLK (pin 7) of the A/D chip.

The six-bit encoded data is output from pins 1 to 6 of the A/D chip and are sent to pins 67 to 72 of the memory controller. The terminals V_{ccA} and V_{ccD} are connected to a $+5$ V supply and the GND terminal is grounded.

6.3.3 Memory, serial/parallel (S/P) and parallel/serial (P/S) conversion

Memory capacity

The capacity of a memory, i.e. the number of memory locations used to store the video signal, is determined by both the sampling frequency (f_{samp}) and the number of quantizing levels (N). In the digital still scheme used by Toshiba, the sampling frequency is 13.3 MHz and six quantizing bits are used. The capacity required to store one field of the video signal is:

$$13.3 \times 10^6 \times 6 \times 0.02 = 1\ 596\ 103\ \text{(bits/field)}$$

The MB81464 is a dynamic random access memory (DRAM) chip used

Analogue and digital special function reproduction **203**

for storage of digital stills. Each chip has a capacity of 60 K × 4 bits, i.e. 64 K (note: 1 K = 1024 or 2^{10}) cells in which four bits can be stored. This means that 262 144 bits can be stored in each chip. Storing of one field of a video signal requires 1 596 103/262 144 = 6.088 65 MB81464 chips. Therefore, six DRAM chips of this type are required to (almost) store one field of a video signal.

Given that one field of a video signal consists of 312.5 lines (two fields in a 625 PAL video signal), it follows that the capacity required for storing one line is 1 596 103/312.5 = 5107.57 bits. Thus, if six DRAMs can store only 262 144 × 6 = 1 572 862 bits, then there is not sufficient storage for 23 239/5107.57 = 4.5 lines of the video signal. When producing a digital still image, 308 lines of data are stored in DRAM in a definite sequence. The lines are written into the DRAM one by one, from the first line to the 308th line, and one after another from the first cell until the 65 536th (64 Kth) cell. The remaining four lines are near the V-BLANK period.

When the lines are read out they are again read out in sequence, from the 1st to the 308th, but cells 32 768 to 33 631 are read out a second time to make up for the lost four lines of the video signal. This is shown in Figure 6.22. Lines 154 to 158 of the video signal are located in cells from 32 768 to 33 631 because the six chips are used in parallel, i.e. after the six cells are filled with 24 bits of data, the address is increased by one. Since each line of data is equivalent to 5107.57 bits and 5107.57/24 = 212.8 address steps, the beginning of lines 154 and 158 will fall into:

the 32 768th cell (32 768/212.8 = 153.98 = 154) and
the 33 631th cell (33 631/212.8 = 158.04 = 158)

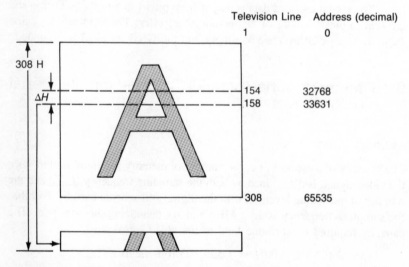

Figure 6.22 Digital still storage video read-out.

Each memory location on the MB81464 DRAM chip can be determined by 16 bits of address (2^{16} = 65 536). The DRAM chip, however, has only 8 pins allocated for the address bus and the microcomputer control system can only provide 8 bits of address at a time. The 16 bits of the address have to be sent in two parts.

There are $\overline{\text{RAS}}$ (row address strobe) and $\overline{\text{CAS}}$ (column address strobe) terminals on the chip. Both are activated by a low logic level which means that 8 bits of the address can be used as a row address. This is taken as the lower 8 bits of the 16-bit address when a low level is applied to the $\overline{\text{RAS}}$ terminal. Similarly, the column address is derived from the higher 8 bits of the 16-bit address when a low logic level is applied to the $\overline{\text{CAS}}$ terminal. Therefore, all the memory cells can be addressed and each cell located by the intersection of the row address and column address.

The DRAM chip has a 4-bit data bus (D_1 to D_4) and the direction of data transportation is determined by the RE/$\overline{\text{WE}}$ control terminal. When the RE/$\overline{\text{WE}}$ terminal receives a low logic level, four bits of data are written into the chosen cell. When the RE/$\overline{\text{WE}}$ is switched to a high logic level, the four bits of data in the chosen cell are sent to the input terminal of a tri-state gate and if the output enable (OE) terminal is also at a low level the data is output from the DRAM chip.

The access time of the MB81464 DRAM used in the DV-98C is 240 ns, but the sampling frequency is $3f_{\text{SC}}$ = 13.3 MHz resulting in a time interval between two adjacent sample points of $1/13.3 \times 10^6$ = 75 ns. The time taken to quantize a sample point pulse, and encode and send the data from the A/D C chip to the DRAM is 75 ns. This means that the speed of processing is three times faster than the access time of the DRAM. The way to get round this problem is to use serial/parallel (S/P) and parallel/serial (P/S) connections.

Serial and parallel conversions

A 4:1 serial/parallel conversion is used for speed matching between the A/D C and DRAM. That is, after all the data from four sample point pulses (24 bits) are determined, they are sent to the DRAM together in parallel. This reduces the speed of data transfer from the A/D C chip to the DRAM chips to $\frac{1}{4}$ of the original speed. The DRAM chip (MB81464) used in the digital still circuit is shown in Figure 6.23.

The S/P conversion scheme used in the DV series of VCR is shown in Figure 6.24. Four sample points, designated a, b, c and d, respectively, are taken as a group. Output of each point of data (6 bits), from the A/D C chip, takes 75 ns so the data from a group of four sample points takes 75×4 = 300 ns (> 240 ns), thus meeting the needs of the DRAMs access speed. In Figure 6.24 the following steps occur:

(a)   the six quantized data of sample point a are sent to the D_1 terminals of the six DRAM chips;

Analogue and digital special function reproduction   **205**

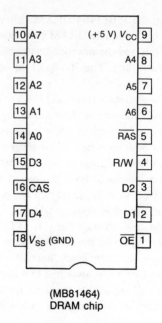

10	A7	(+5 V) V_{CC}	9
11	A3	A4	8
12	A2	A5	7
13	A1	A6	6
14	A0	$\overline{RAS}$	5
15	D3	R/W	4
16	$\overline{CAS}$	D2	3
17	D4	D1	2
18	V_{SS} (GND)	$\overline{OE}$	1

(MB81464)
DRAM chip

Figure 6.23 DRAM chip used in the digital still circuit.

(b) the six quantized data of sample point b are sent to the D_2 terminals of the six DRAM chips;

(c) the six quantized data of sample point c are sent to the D_3 terminals of the six DRAM chips; and

(d) the six quantized data of sample point d are sent to the D_4 terminals of the six DRAM chips.

Thus, the data of a group of sample points (four points), namely, D_{a0}, D_{b0}, D_{c0} and D_{d0}, are located in one cell of the ICV04 chip. Similarly:

D_{a1}, D_{b1}, D_{c1} and D_{d1} are located in ICV05
D_{a2}, D_{b2}, D_{c2} and D_{d2} are located in ICV06
D_{a3}, D_{b3}, D_{c3} and D_{d3} are located in ICV07
D_{a4}, D_{b4}, D_{c4} and D_{d4} are located in ICV08
D_{a5}, D_{b5}, D_{c5} and D_{d5} are located in ICV09

All the (24-bit) data of a group of (four) sample points are stored in the same cell in each of the six chips which have the same address. The S/P converter has a conversion ratio of 6:24 and is manufactured as a large scale integrated chip (LSI), type T671147, memory controller.

Parallel/serial (P/S) conversion is the inverse of serial/parallel (S/P) conversion, converting the 24-bit data read-out from the six DRAM chips into six serial data. When 24-bit parallel data is output from the DRAM chips,

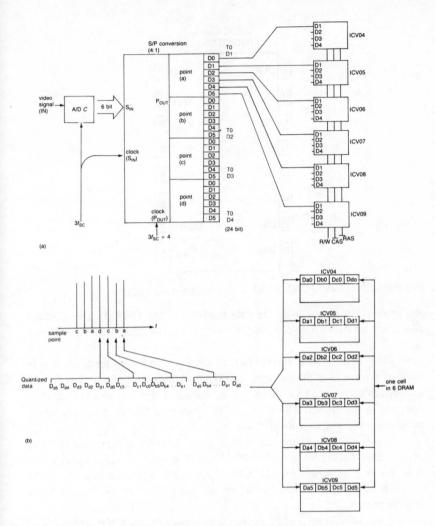

Figure 6.24 S/P conversion in the DV series VCR.

each of the six input terminals of the D/A C chips receives 4-bit serial data, reducing the speed of operation of the DRAM chips to $\frac{1}{4}$ of the original access speed. The parallel-in terminals can use a clock frequency of $\frac{3}{4}f_{SC}$, while the serial-out terminals use $3f_{SC}$. Like the S/P converter, the P/S converter is integrated in the T671147 memory controller chip.

6.3.4 D/A conversion

The D/A conversion is the inverse of A/D conversion and recovers the original

Analogue and digital special function reproduction **207**

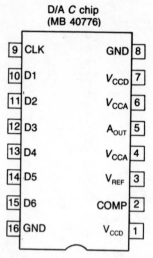

D/A C chip
(MB 40776)

9	CLK	GND	8
10	D1	V_{CCD}	7
11	D2	V_{CCA}	6
12	D3	A_{OUT}	5
13	D4	V_{CCA}	4
14	D5	V_{REF}	3
15	D6	COMP	2
16	GND	V_{CCD}	1

Figure 6.25 D/A C chip used in the digital still circuit.

analogue video signal from the data read-out of the DRAM. The terminal allocations for the D/A chip, MB40776, used in the DV-98C, is shown in Figure 6.25.

On both the A/D C and the D/A C, the V_{CCA} and V_{CCD} terminals denote the source voltages of the analogue and digital sections respectively. The COMP terminal is connected to a compensating capacitor to improve the frequency response of the circuit. The V_{ref} terminal requires a reference voltage which determines the output voltage from the D/A C.

Figure 6.26 shows the circuit from the D/A C chip to the output terminal of the digital still circuit in the DV-98C VCR. The analogue video signal output from the A_{out} terminal of the D/A C chip is sent to the open terminal of the switch ICF02 via V26. The LPF and the two buffers (QV25, QV24) are shown in Figure 6.26.

The low pass filter (LPF) consists of LV03, 04 and CV10, 11, 12, and filter the staircase ripple (frequency $3f_{SC}$) generated in the D/A conversion. The output terminal of QV25 is connected to LV02 and CV09 to improve the high frequency characteristic. A larger capacitor, CV13 (CV13 $\gg$ CV12), is added to the output terminal of QV26 and is controlled by QV27.

A positive pulse is sent from the H-SYNC terminal of the memory controller in every field, and is used to turn on QV27 so that the output video signal from QV26 is reduced and the chroma signal is eliminated by the ACK in the monitor. The pulse is 260 μs in width, a little bit wider than four lines of video and appears in such a position to just cover the additional four lines. Moreover, because it appears at the end of one field, quite near to the V-SYNC signal, it can be used as a pseudo V-SYNC signal to replace the PB V-SYNC signal in the still mode.

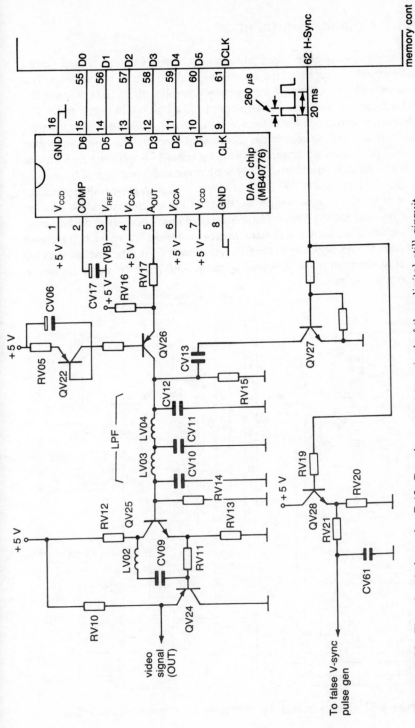

Figure 6.26 The circuit from the D/A C to the output terminal of the digital still circuit. Note: V_B is the same as that in Figure 6.21.

6.3.5 Memory controller

The memory controller is central to the operation of the digital still circuit and provides the data bus, the address bus and the control bus sections. The function and pin outline for the memory controller chip, T671147, and the I/O interface have already been discussed and are shown in Figure 6.27. The address bus section consists of the address generator, while the clock generator, the V counter and the control bus section are formed by the timing controller.

The address generator has two 8-bit counters to generate an 8-bit row and an 8-bit column address. The row address counter is triggered by the lock at $3f_{SC}$, adding 1 every 300 ns up to a count of 256. A pulse is then generated to reset the row address counter and then add 1 to the column address counter. The row and address counters have to be reset simultaneously at the beginning of each field to ensure that the data written in, or read out, of the first memory location always begins with the first line of each field. The two counters are reset by the reset pulse generated by the K REF pulse via the V counter as shown in Figure 6.28.

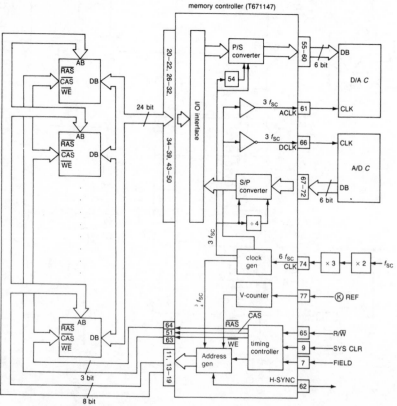

Figure 6.27 The pins and the functions of the memory controller.

The K REF pulse is actually the frame reference ($\frac{1}{2} V_D$) signal, which is a symmetrical square wave with a period of 40 ms and a leading edge near the V-SYNC pulse. Its period is reduced to 20 ms (that of the field period) by a frequency doubler and its waveform is that shown in Figure 6.28b. The signal is then delayed via the monostable MM1 to a position just 260 μs ahead of the end of each field so that the output from the $\overline{Q}$ terminal of MM1 can trigger another monostable, MM2, to output a 260-μs-wide positive pulse at the last four lines of each field and act as a reset pulse for the column address counter.

Even when all 65 536 memory locations have been read out and the row and column addresses are 1, there is still the time period equivalent to four lines to elapse before the reset pulse is generated. So lines from 154 to 158, corresponding to the addresses from 32 768 to 33 631 (1000000000000000

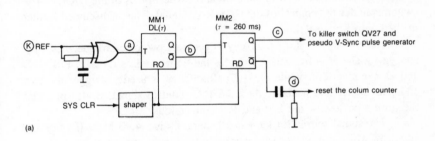

(a)

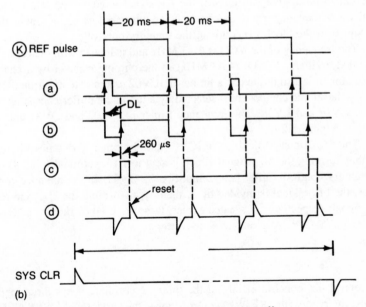

(b)

Figure 6.28 The V counter in the memory controller.

to 1000001101011111) have to be read out again. This means that when the row and column address counters reach 1, the row address counter is reset and the column address counter is preset to 1000000000000000. The $3f_{SC}$ clock continues to trigger the row address counter to add 1 every 300 ns until it reaches 256. The row counter is reset and the column counter is incremented by 1. The process is repeated until the column address counter reaches 10000011 and the row address counter is 01011111. The reset pulse then resets the two counters together to ensure that the first line of each field is either read out or written in the first memory location.

It can be seen from Figure 6.28 that the 260-μs-wide pulse appears just at the last four additional lines. It is sent from the memory controller to the switch Q27 and used to form a pseudo V-SYNC pulse as described previously. The 8-bit address output from the row and column address counters are selected by an 8-bit switch and sent to the 8-bit address terminals of the DRAM. The switch is under the control of the CAS/$\overline{\text{RAS}}$ signal. This is discussed further in the control bus section.

The clock generator needs only to generate two clock frequencies, $3f_{SC}$ and $\frac{3}{4}f_{SC}$. The $3f_{SC}$ frequency provides the timing pulses for S/P, P/S, A/D and D/A converters, while the $\frac{3}{4}f_{SC}$ frequency is used to trigger the row address counter and generate the CAS/$\overline{\text{RAS}}$ control signal. It also provides the timing pulses for the S/P and P/S conversions.

The signal source for the clock generators is a 4.43 MHz (f_{SC}) crystal oscillator which is used by a subcarrier converter in the chroma channel. Having been sent to the digital still circuit, the f_{SC} signal is converted into a $6f_{SC}$ signal via a frequency tripler and frequency doubler before being sent as the $6f_{SC}$ signal to the clock generator in the memory controller.

The frequency of the VCO is 4.43 MHz and is locked with the crystal-controlled oscillator (VXO) (4.43 MHz) in the chroma channel by a phase comparator. It is also limited by a limiter in ICV02 to generate its harmonics. The $3f_{SC}$ harmonic component is selected by a tuning amplifier consisting of QV31, CV29 and LV08. The frequency tripler is ICV02 and QV31 and is shown in Figure 6.29.

The $3f_{SC}$ signal is doubled in ICV03 and the other harmonics filtered out, thus forming the $6f_{SC}$ signal which is sent to CLK terminal (pin 74) of the memory controller after being amplified by QV32. In the memory controller, the $6f_{SC}$ signal is divided by 2, then by 4, forming the $3f_{SC}$ and the $\frac{3}{4}f_{SC}$ signals respectively. As stated earlier, these are used by the two address counters, P/S, S/P, A/D and D/A converters.

Control bus

The control bus consists mainly of the timing controller. It has three input signals; the read/write (RE/$\overline{\text{WE}}$) control signal, the system clear (SYS CLR) signal and the field synchronizing (FIELD) signal. The $\frac{3}{4}f_{SC}$ signal, formed

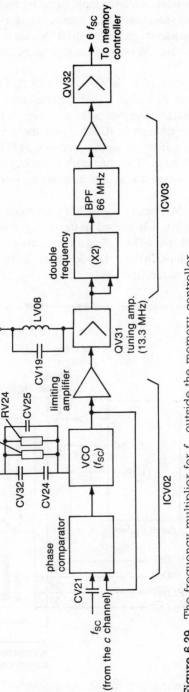

Figure 6.29 The frequency multiplier for f_{sc} outside the memory controller.

Analogue and digital special function reproduction **213**

in the memory controller, is also an input signal for the timing controller. The timing controller provides the counter's power-up reset signal, the CAS/RAS control signal for the address generator, the R/W control signal, the row address strobe signal ($\overline{\text{RAS}}$) and the column address strobe signal ($\overline{\text{CAS}}$) for the six DRAM chips.

The three input signals (RE/$\overline{\text{WE}}$, SYS CLR, FIELD) are from logic control circuits while the K REF signal is derived from the servo system. The K REF signal is divided into two paths as shown in Figure 6.30. One path is to the input mode detector in IC601. When the still mode is selected, the three signals are sent from the output processor in IC601 at the first trailing edge of the K REF pulse. The other path is directly to the V counter in the memory controller where it is used to generate the reset pulse for the column address counter.

After the RE/$\overline{\text{WE}}$ control signal has been received at the memory controller through pin 10, it is latched in the timing controller and transferred to the WE terminal (pin 63) by the column address counter reset pulse and sent to the DRAM chips. In the still mode, all the field is read out from memory with the exception of the first field.

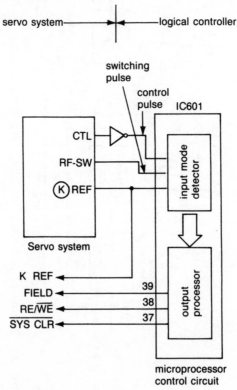

Figure 6.30 Input signals in the timing controller.

The SYS CLK signal is sent to the memory controller through pin 9. It is shaped in the timing controller and then distributed to the counters in the address generator, the monostable in the V counter, and the latch in the S/P converter, to reset them at the beginning of the still mode.

The FIELD signal has been replaced by the K REF pulse in new models of the machine.

The RAS, CAS and CAS/$\overline{\text{RAS}}$ signals generated by the timing controller are formed from the $\frac{3}{4}f_{SC}$ signal as illustrated in Figure 6.31. Writing in, or reading out, one location of the six DRAM is completed within 300 ns. During this period, there is a need to send the row or column address separately, and send low logic levels to the CAS or RAS terminals correspondingly. In Figure 6.31 the row address and the column address are generated from their own counters and sent to an 8-bit switch which is controlled by the CAS/$\overline{\text{RAS}}$ control signal.

The CAS/$\overline{\text{RAS}}$ control signal is output from a monostable which has a $\tau > 150$ ns, which is triggered by the $\frac{3}{4}f_{SC}$ signal. First, the monostable outputs a high logic level, selecting the 8-bit column address and keeping the CAS terminals of the DRAM at a low logic level. Then, the monostable outputs a low logic level, selecting the 8-bit row address and keeping the RAS terminals at a low logic level after the period $\tau > 150$ ns has elapsed. This ensures that the column address and the row address are sent in a definite time sequence. This must be completed within 300 ns, so the operating time of the monostable is required to be less than 150 ns.

6.3.6 Digital slow motion technique

The digital slow motion technique uses the field memory to obtain noise-free slow motion PB pictures at $\frac{1}{16}$ of normal tape speed for the DV-90D/DC and DV-98C, and $\frac{1}{3}$ of normal tape speed for the DV-93D/DC and DV-94C. The $\frac{1}{16}$ of speed slow motion will be used as an example for discussion purposes.

Figure 6.32 shows the capstan servo circuit for the DV-90D/DC and DV-98C. In the normal PB mode, the capstan is driven by the capstan speed loop and the capstan phase loop ($\times$ 1). The speed loop, which is referred to as the 'discriminator' in Figure 6.32, detects the FG pulse of the capstan motor using a digital circuit, i.e. PWM. The phase loop, which is also called the 'phase comparator' in Figure 6.32, compares two signals (the frame reference signal, $\frac{1}{2}V_D$ and the PB CTL signal) using pulse width modulation (PWM). The frame reference signal is generated by the field oscillator (V_{OSC}) and sent to the PWM via the tracking monostable in IC501. Because the switch IC504(c) is connected to the closed position in the normal PB mode ($\times$ 1), the tracking control is dominated by R553 and R556. (R556 is located on the front panel and is labelled as tracking control.)

Three other switches (IC504(a), (b) and (d)) are also connected to the

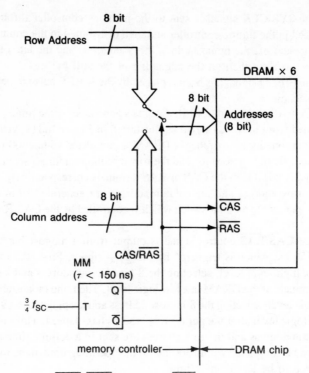

Figure 6.31 The $\overline{RAS}$, $\overline{CAS}$ and CAS/$\overline{RAS}$ signals generated in the memory controller.

closed terminal in the normal playback mode ($\times$ 1), so the output of the capstan speed loop is mixed with that of the capstan phase loop ($\times$ 1). No negative feedback is added to the amplifier IC503. Hence, the capstan motor rotates at normal speed and the speed detection signal sent to the microprocessor control circuit is the PB CTL signal.

When the SLOW key is depressed, the slow signal, a low logic level, is output from IC601-28 of the microprocessor control circuit. After inverting to become the slow(H) signal it is distributed to the four switches of IC504, as shown in Figure 6.32. IC504(a) is closed and negative feedback is added to the amplifier IC503 and reduces the rotating speed of the capstan motor, and reduces the frequency of the PB CTL pulse.

The digital servo has to handle the new FG pulse. The switch, IC504(b), changes from the closed to the open terminal and the output of the capstan phase loop is mixed with the capstan speed loop ($\times \frac{1}{6}$). The digital servo circuit for the capstan phase loop is integrated in IC507, and is known as the phase comparator. It compares the phase of the PB CTL signal with that of the frame reference signal, and outputs an error voltage which is proportional to the difference between them. However, the frequency of the PB CTL signal

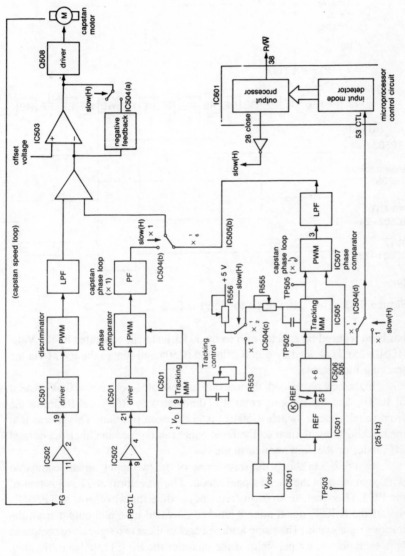

Figure 6.32 The capstan servo circuit in DV-90D/DC, DV-98C.

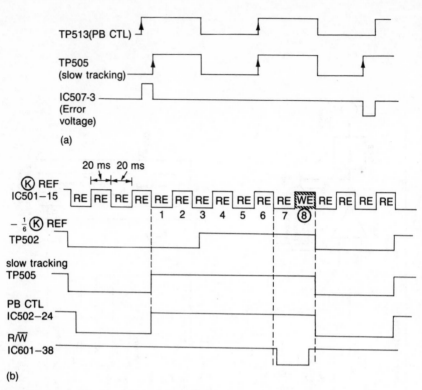

Figure 6.33 The waveforms for Figure 6.32.

has been reduced by a factor of 6 so the K REF signal, generated by the V_{osc} (IC501-25), is divided by 6 in IC506 and IC505 and sent to the PWM via the tracking monostable.

In the new phase loop, the tracking monostable is located in IC505 instead of IC501, and the tracking control is dominated by R555 and R556 on the front panel because switch IC504(c) is in the open position. This means that the tracking control knob on the front panel can be used in either the normal PB mode, or the slow motion mode.

Figure 6.33a shows the waveforms of the two input signals compared with each other in the capstan phase loop. The waveform at TP513 is that of the PB CTL signal, reduced in frequency, while the waveform at TP505 is that of the K REF signal after it has been divided by 6 and output from the tracking monostable. The rising leading edges of these two signals are compared with each other once per field. If the phase of the PB CTL pulse leads, then IC507-3 will output a positive pulse, the width of which is determined by the magnitude of the phase difference. If the phase of the PB CTL pulse lags behind the divided K REF pulse IC507-3 produces a negative pulse whose width also depends on the phase difference. These pulses are passed through a low pass filter and are converted to either a negative or positive d.c. voltage depending on the original pulse polarity. The magnitude of the d.c. voltage depends on the pulse width.

The tracking control moves the rising edge, at TP505, forwards or backwards to change the phase difference and the d.c. error voltage output from the LPF. The rotational speed of the capstan motor, and the tape speed, will vary to adjust the scan path of the video heads to that of the recorded tracks.

Figure 6.33b illustrates how to select the write-in or read-out mode for each field of the K REF signal. The waveform at TP502 is that of the $\frac{1}{6}$ K REF pulse, the period of which is six times that of the K REF pulse. While the capstan motor rotates at $\frac{1}{6}$ normal speed, the period of the PB CTL is also extended by a factor of six. This means that it will take the time period for six frames to cover a one-frame distance, i.e. two track intervals.

The picked up PB signal is a maximum when the scan path of the video head follows the recorded video track at the eighth field after the rising edge of the PB CTL pulse. Therefore, the PB signal of this field is selected to be written into the memory and the other fields of the six frames are read out. That is, one field is written into memory and this field is read out repeatedly for eleven field periods, every six frames. The operation is carried out by the microprocessor control circuit (IC601-53) which starts counting when it receives the trailing edge of the $\frac{1}{6}$ K REF pulse and stops when it reaches the eleventh period of the K REF pulse. A low logic level, i.e. the next WE signal, is output from IC601-38 when the counter stops. The process is repeated when the next trailing edge of the K REF pulse is detected. It can be seen that the operation of both the digital still and slow motion circuits are the same except for the RE/$\overline{\text{WE}}$ signal.

Chapter 7

Standard and high band 8 mm VCR

On 26 April 1983 a conference of 127 companies world-wide published a document proposing an 8 mm VCR format standard. The standard described five aspects relating to video, audio, tracking, cassette and magnetic tape. These five features form the basis of the now accepted 8 mm VCR standard.

High band 8 mm VCR is an option of the standard 8 mm format in which not only the quality of the PB picture, such as resolution and S/N ratio, has been improved but also features such as time-code have been adapted. Editing in the U-matic format was also further developed. The first 8 mm standard VCR was released by the Sony Company in January 1985 followed by a high band 8 mm VCR in April 1989.

A number of new techniques particular to the 8 mm VCR will be introduced in this chapter. These include aspects relating to the video, audio and tracking as well as descriptions of the metal tape and TSS (tilting sendust spatting) head.

7.1 Metal tape and the tilting sendust spatting (TSS) head

The metal tape used in the 8 mm VCR was developed by TDK in 1984. Only MP (polished) tape was used in the standard 8 mm VCR. The MP tape is coated with super-fine magnetic particles of cobalt and nickel, with a particle size of about 0.2 μm. Its BET area, defined as the total surface area per gram of magnetic particles, is not less than 55 m^2/g, which is about 1.5 times greater than that of magnetic oxide-particle tape (35 m^2/g).

The magnetic oxide-particle coating on oxide tapes contains a lot of non-magnetic oxide ions. In the metal tape coating, these oxide ions are replaced

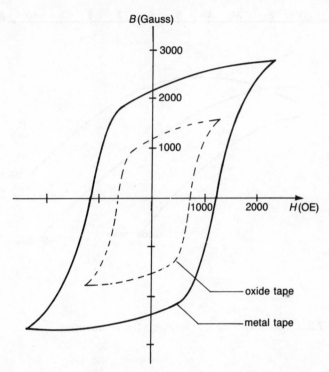

Figure 7.1 Comparison between the metal and oxide tapes.

by ferric ions in an oxide film several nanometres thick. These features greatly improve the magnetic performance of the tape. Figure 7.1 shows a comparison between the magnetic performance of metal and oxide tapes.

In addition to new coatings, a new binder is used on 8 mm metal tapes. The binder is used to ensure adhesion of the magnetic particles to form an even film. The new binder consists of two binders and several additives. One of the binders ensures an even diffusion of the magnetic particles, while the other is used as a fast adhesive to stick the particles to the tape base. The additives are used to provide a low temperature coefficient, an appropriate friction coefficient and good lubrication. These features mean that the surface coating and its adhesion have been greatly improved.

During the manufacturing process, a mirror polish technique is applied to the tape to ensure a smooth surface finish to better than 0.2 μm. The whole manufacturing process, including preparing material, plating and evaporation, cutting out and assembling is carried out in a dust-free environment, ensuring tape manufacture with the best possible surface coating.

A new metal tape, the ME (evaporated) tape, is used for the high band 8 mm VCR. The ME tape has an evaporated coating of super-fine magnetic particles of cobalt and nickel. The new vacuum evaporation plating technology

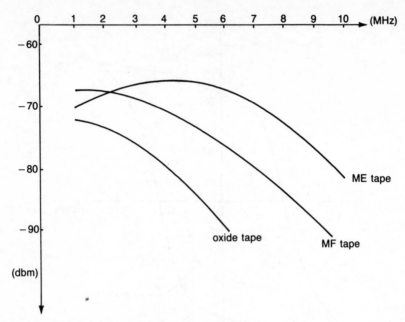

Figure 7.2   The output characteristic of various tapes.

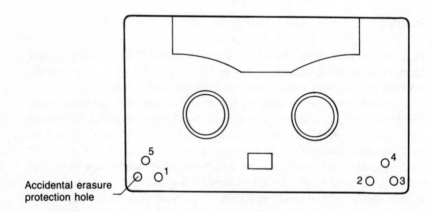

Figure 7.3   Recognition hole for MP and ME tapes. Note:

	hole 1	hole5
standard MP	closed 0	closed 0
high band MP	closed 0	open 0
high band ME	open 0	closed 0

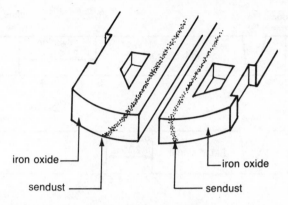

iron oxide

iron oxide

sendust

sendust

Figure 7.4 Configuration of the TSS heads.

means that the tape can be made without any binder and results in an increased magnetic density (*Br*). The magnetic characteristic of the metal tape is superior to that of the oxide, and the metal tape more easily saturates the scanning head.

A comparison between the output level of an ME and MP tape is shown in Figure 7.2. Recognition of the MP and high band MP and ME tapes is determined using holes 1 and 5 in the cassette as shown in Figure 7.3.

A complex head, referred to as the sendust head, is made by tipping the ferric oxide head with a higher saturation level material known as sendust. The sendust heads have been used on 2-inch tape VTRs, which are used for broadcast quality applications, until it was discovered that a non-magnetic layer builds up on the surface of the head tip after scanning the oxide tape. This has the effect of reducing the quality of playback.

A new TSS head has been developed for use with the 8 mm VCR. Its configuration is shown in Figure 7.4. Narrowing of the gap in the sendust head has improved the PB frequency response, and combining the mono and multicrystalline iron oxides has reduced friction noise generated on the iron oxide surface of the TSS head.

7.2 Improvements in the video channel

A number of improvements in the video channel of the 8 mm VCR have been made.

A shorter wavelength

If a metal tape and a TSS video head are used, the record wavelength (λ) of 0.4 μm is possible. This is the shortest wavelength (highest frequency) in all PAL system VTRs. The short wavelength permits a shift in the FM *Y* signal

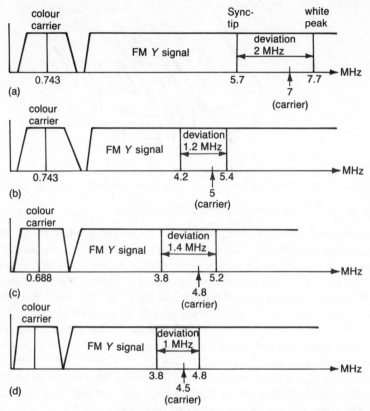

Figure 7.5 The carriers and the deviations of FM Y signals for 8 mm and other VCRs: (a) high band 8 mm, (b) standard 8 mm, (c) Betamax, (d) VHS.

carrier from 4.5 MHz for VHS (4.8 MHz for Betamax) to 5 MHz for standard 8 mm and 7 MHz for high band 8 mm VCRs. This means that the horizontal resolution for 8 mm high band VCRs can be greater than 400 television lines, as shown in Figure 7.5. It can also be seen from Figure 7.5, that the deviation of the FM Y signal, for the 8 mm VCR, is wider than that of Betamax, VHS and standard 8 mm VCR, making it possible to reduce the noise figure of the PB signal. In addition, narrowing of the gap of the TSS head produces an increase of 2 dB at 7 MHz in the output of the head.

Increased high frequency signal

The emphasizer used in the 8 mm VCR, shown in Figure 7.6, comprises a main emphasizer and subemphasizer. The subemphasizer serves as a non-linear circuit, while the main emphasizer is employed as a linear emphasizer and

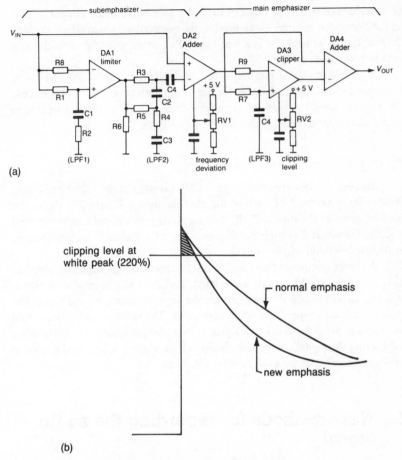

Figure 7.6 The emphasizer in 8 mm VCR: (a) the emphasizer in 8 mm VCR, (b) the improvement using the emphasizer in high band 8 mm VCR.

white/black clipper. In the subemphasizer, the low frequency component is applied to the negative terminal of an adder (DA2) through LPF1, LPF2 and limiter (DA1), without limiting if the signal is small. The result is that the high frequency components of the video signal, which is output from the differential amplifier (DA2), are greatly increased.

The gain of DA2 is controlled by RV1. This is referred to as the frequency deviation control, because the output level from the subemphasizer will determine the frequency deviation of the frequency modulator. The low frequency component amplitude is limited by the limiter (DA1) when the level of the input signal is high to maintain non-linear emphasis.

The operating principle of the main emphasizer is analogous to that of the subemphasizer except that DA3 is used as a white/black clipper to clip out the overshoot sections as the level of the emphasized video signal exceeds 220% at the white peak, or is lower than 100% at the sync-tip. The clipping level is controlled by RV2.

It should be noted that the time constant produced by R7 and C4 is shortened from 1.3 μs to 0.47 μs for the 8 mm high band VCR. This has the effect of reducing the noise at the edge of the picture to about $\frac{1}{3}$ of that in the standard 8 mm VCR. This is shown in Figure 7.6b.

Improved signal-to-noise ratio

A digital chroma noise reducer (digital CNR) is used in the high band 8 mm VCR to improve the S/N ratio of the chroma signal. Figure 7.7 shows the block diagram of the digital CNR. The digitized video signal, including both signal and noise, is divided into two paths. One of the paths is used to generate an inverted version of the noise in the signal.

A direct component and an inverted delayed component are added in this path. Because the signals are related, and the noise is random between adjacent lines, the added result cancels the signal component and leaves the noise components superimposed on each other. The resultant noise is inverted and added to the delayed original signal in the other path, suppressing the noise and leaving the signal component. Noise which is amplitude modulated on to the signal can be suppressed by 4−5 dB in this way.

7.3 New methods for recording the audio signal

In the traditional VCR, the method used to record the audio signal is the same as that in the audio recorder, i.e. using a fixed audio head, remaining on one (or two) longitudinal track(s). As the tape speed is reduced to prolong the recording and replay time, the qualities of the replaying audio signal, such as frequency response, S/N ratio and flutter and wow will deteriorate. The method used to solve these problems is to record the audio signal using rotating video heads in the same manner as the video signal. The scanning speed of the audio track is still 6 m/s although the tape speed is less than this when still or slow modes are selected.

Two methods have been used to record the audio signal. Both FM and PCM have been utilized in the 8 mm VCR and other hi-fi VCR systems. For a hi-fi VCR, the rotating heads recording an audio FM signal scans the 180° tape wrap in just the same way as the video heads. They share the scan with the video heads for hi-fi Betamax or 8 mm VCRs. The major difference between the latter two is that the left and right hand audio signal is frequency

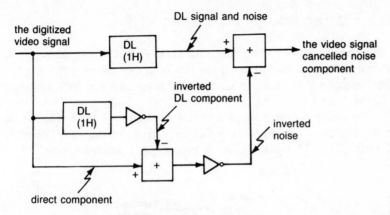

Figure 7.7 The block diagram of digital CNR.

modulated on to 1.4 MHz and 2.1 MHz, with a deviation of ± 500 kHz for the hi-fi Betamax, while there is only one audio carrier of 1.5 MHz ± 100 kHz for the 8 mm VCR.

The tape has to wrap 221° to record both the video and audio signal when the audio is encoded using PCM. The audio track occupies 31° and is located outside the 180° used for the video track. This is shown in Figure 7.8. When the audio track uses FM, it is not possible to dub audio after the video signal is recorded because of the shared nature of the signal.

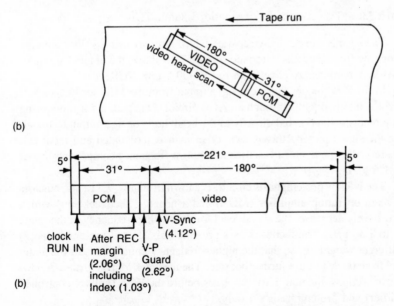

Figure 7.8 Tape format for recording audio PCM signal: (a) tape format, (b) area allocation.

7.3.1 FM recording in the 8 mm VCR

As described previously, the track of the audio FM (AFM) signal is shared with that of the modulated video signal. However, they occupy different parts of the frequency spectrum. An example of this for an 8 mm VCR is shown in Figure 7.9.

The carrier for the *Y* signal is from 4.2 MHz to 5.4 MHz. The carrier for the *C* signal is 0.743 MHz, while the carrier for the AFM signal is 1.5 MHz. The ATF signal, shown in Figure 7.9, is discussed later.

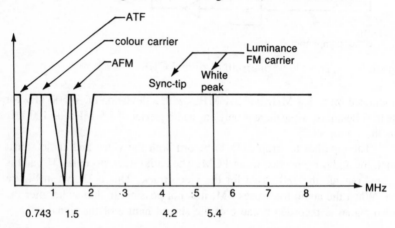

Figure 7.9   The spectrum for standard 8 mm VCR.

The signals can be easily separated during replay using a filter with the appropriate frequency characteristic. Figure 7.10 shows the block diagram of the AFM record/replay channel in the V-900 8 mm VCR.

In the REC mode, the audio signal input from the MIC, or line socket, is divided into two paths through an AGC circuit. One path is for monitoring during recording and is fed directly to the headphone or line output terminal, while the signal passing through the other path is modulated and recorded. In the recording path, an LPF is used to suppress frequency components higher than 1.5 MHz.

The NR, or pre-emphasis circuit, is shown in Figure 7.11a and consists of a main operating amplifier (MOA) and a negative feedback loop which progressively decreases the signal level for decreasing frequency in the range 2.12 to 8.38 kHz. The feedback uses pre-emphasis (1) and pre-emphasis (2) of different weighting, so that the higher frequency components of the audio signal from the MOA are further boosted. The feedback loop is further divided into two sections, the non-linear emphasis before the VCA (voltage controlled amplifier) and pre-emphasis (1) only.

The gain of the VCA is controlled by the output level from the detector and weighting filter. This means that, as the VCA is controlled by the filtered

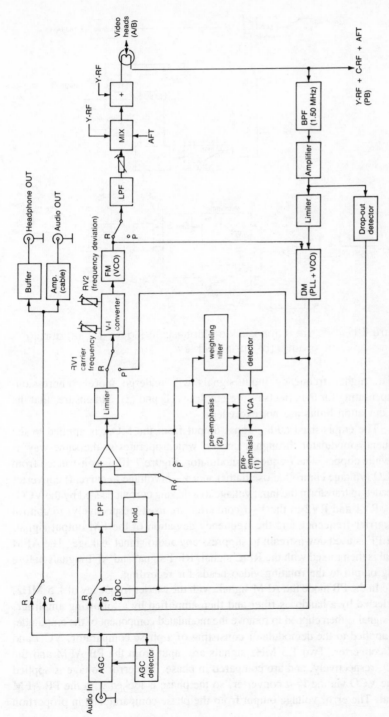

Figure 7.10 Hi-fi AFM record/replay in V-900 8 mm VCR.

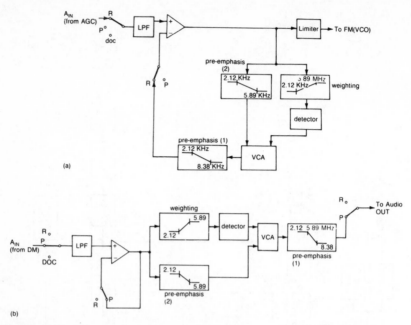

(a)

(b)

Figure 7.11 NR pre-emphasis/de-emphasis in Figure 7.8: (a) during recording, (b) during replay.

signal, higher frequency input signals will undergo more compression, compensating for the effects of pre-emphasis (1) and (2), and ensures that the FM deviation limits are not exceeded.

The emphasized audio signal output from the MOA is applied to the frequency modulator through a limiter, which operates in the same way as the white clipper. The frequency modulator, Figure 7.10, is constructed from a VCO (voltage controlled oscillator) so a $V-I$ (voltage to current) converter is needed to transform the input voltage to a driving current needed by the VCO.

RV1 and RV2 in the $V-I$ converter are used, respectively, to control the carrier frequency and the frequency deviation of the FM output signal. An LPF is used downstream to suppress any audio signal leakage. The AFM signal is then mixed with the RF C signal, RF Y signal and AFT signals before being output to the rotating video heads for recording.

In the PB mode the AFM signal, with the carrier frequency of 1.5 MHz, is selected by a bandpass filter and then amplified by a following amplifier. The signal is then clipped to remove the modulated component of the amplitude, and applied to the demodulator consisting of a phase comparator, VCO and $V-I$ converter. Two 1.5 MHz signals are input from the PB AFM and the VCO, respectively, and are compared in phase. The error voltage is applied to the VCO via the $V-I$ converter, so the phase is locked with the PB AFM signal. The error voltage output from the phase comparator is in proportion

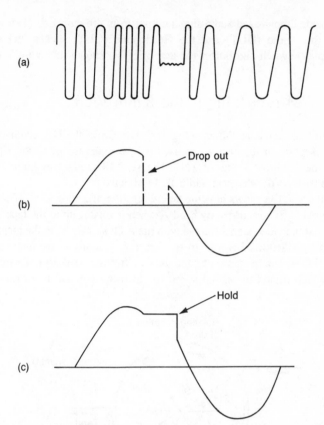

(a)

(b)

Drop out

(c)

Hold

Figure 7.12   The waveforms for drop-out compensation: (a) drop-out in
AFM signal, (b) drop-out in audio signal demodulated,
(c) drop-out section is replaced by hold level.

to the frequency of the PB AFM signal and is thereby the demodulated audio
signal. An LPF is used downstream to suppress any carrier leakage.

When a drop-out occurs the result is a reduction in the level of PB audio
FM carrier and a drop-out pulse is generated by the drop-out detector. This
pulse is applied to the switch to select the DOC and also to the hold circuit.
This means that the dropped out section in the demodulated audio signal can
be replaced by a hold level, which equates to about 75% peak audio level.
This is shown in Figure 7.12.

The NR or de-emphasis circuit is located after the demodulator (DM)
as shown in Figure 7.11b. The frequency-tailored pre-emphasis circuit is not
inserted in the feedback loop, but in the direct path. The effect of this is to
reverse the operation of the circuit to become an expander whose amplitude
and frequency response mirrors that introduced in the record case. The tape
noise can be attenuated during pre-emphasizing and de-emphasizing.

The de-emphasized audio signal output from pre-emphasis (1) is output to both the Audio (OUT) and the Headphone (OUT) sockets via a R/P (record/play) switch and buffer (or amplifier) as shown in Figure 7.10.

7.3.2 PCM recording in the 8 mm VCR

The audio PCM track, as shown in Figure 7.8, occupies the 31° section outside the 180° section of the head wrap used by the video track, so the tape has to wrap the drum for a total of 221°. Figure 7.13 shows how both sections are used to record video and audio PCM signals.

Fugure 7.13a shows how the tape wraps the drum for 221°. The audio PCM signal is first recorded by head A when it rotates up to the tape. It can be seen that the video signal (mixed with the ATF signal) is, in the meantime, recorded by video head B (one of two ways for recording the audio signal, AFM or PCM, can be chosen by the user). After the head drum rotates 31°, during which period the signal is fed to video head A and recorded on the

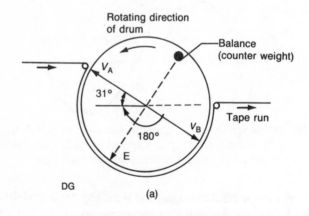

(a)

(during recording)

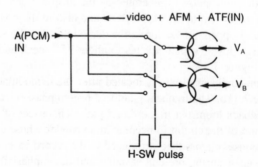

(during replay)

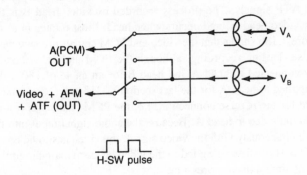

A(PCM)
OUT

Video + AFM
+ ATF (OUT)

V_A

V_B

H-SW pulse

(b)

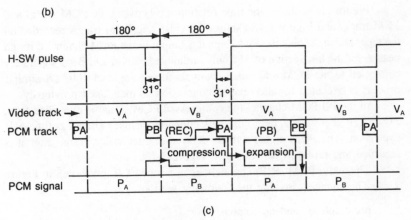

H-SW pulse

180°   180°

31°   31°

Video track → V_A V_B V_A V_B V_A

PCM track   PA   PB (REC)→ PA   (PB)   PB   PA

compression   expansion

PCM signal P_A P_B P_A P_B

(c)

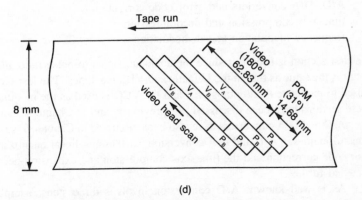

Tape run

8 mm

Video
(180°)
62.83 mm

PCM
(31°)
14.68 mm

video head scan

V_B V_A V_B V_A

P_B P_A P_B P_A

(d)

Figure 7.13 PCM audio and video recording. (a) the tape wrapping
the drum for 221° (V_A, V_B = video heads A, B
respectively; E = erase head), (b) the video and PCM
signals are switched by the H-SW pulse, (c) the
relationship between PCM signal and PCM track, (d) the
tracks recorded consist of the video and PCM tracks.

Standard and high band 8 mm VCR   **233**

tape, switching from the H-SW pulse occurs and the signal is replaced by the video and ATF signals. The process recorded by video head B is the same as that for video head A, and occurs when head B just rotates in to the tape.

Figure 7.13b shows that the video and PCM signals are switched by the H-SW pulse. During recording, although the PCM and video channels are separately connected to head A and head B for an angle of 180°, the PCM signal is applied to head A for the last section of 31° only. A similar process is executed for the reverse connection, i.e. the PCM channel is connected to head B and the video to head A. Because the audio signal input into the VCR appears simultaneously with the video signal, the video has to be compressed from 180° to 31° before being fed to the video heads to maintain timing. This is known as time-axis compression.

Figure 7.13c shows the time relationship between the PCM signal and PCM track, and Figure 7.13d shows the video and PCM tracks recorded on the tape. During replay the PCM signal is output to the PCM channel through head A for the last section of 31° only, although heads A and B are separately connected to the PCM and video channels for an angle of 180°. A similar process is executed for the reverse connection as mentioned previously.

In the PB PCM channel the compressed PCM signal is expanded from 31° to 180° before it is fed to the D/A converter. This is known as time-axis expansion. The PCM signal will be delayed one television frame after it is recorded and replayed, as shown in Figure 7.13c.

The audio PCM (R/P) channels in an 8 mm VCR are shown in Figure 7.14. The channels involve the following sections:

pre-emphasis and de-emphasis
A/D, D/A converters and error code correction
time-axis compression and decompression
frequency modulation and demodulation

The first section is used to reduce the analogue signal-to-noise ratio and its circuit is the same as that in the U-matic and VHS machines. The last section is used to modulate the digital signal and the VCO is used as the modulator. These sections have been discussed just previously and in Chapters 3 and 5.

A/D and D/A converters have also been discussed in Chapter 6, so only some difficulties in A/D and D/A conversion, such as non-linear quantization, error code correction and the time-axis compression and expansion need be discussed further.

As is well known, A/D conversion involves three steps; sampling, quantizing and encoding. The sampling frequency is $2f_H$, i.e. 31.25 kHz, in the 8 mm VCR. Because the audio information on the tape has to be compressed from an angle of 180° to 31°, the recording data has to be compressed to 8 bits per sampled value. A 10−8-bit converter, a non-linear quantizer, is used to improve the digital dynamic range.

In quantization theory, the digital dynamic range is defined as the ratio

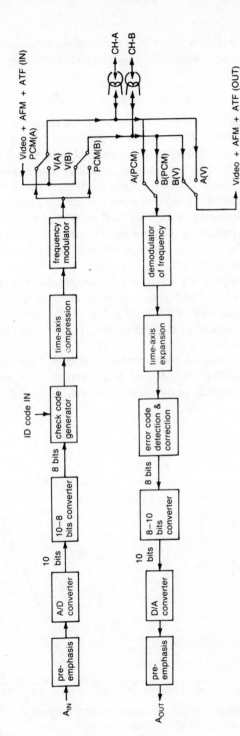

Figure 7.14 Audio PCM (R/P) channel.

of quantization error to the maximum of variable range of the signal, and the dynamic range for linear quantization is only determined by the number of quantization bits, i.e.

dynamic range $= 6 \times$ number of bits $+ 1.8$ (dB)

A wider dynamic range can be achieved using non-linear quantization, although only with the same number of bits used in the equivalent linear quantization.

Non-linear and linear quantizations, respectively, correspond to floating point and fixed point expressions of a number. A floating point number involves a mantissa and an exponent, which is a variable. If three decimal digits are used for encoding quantization levels, the maximum variable range of the signal using fixed point representation is from 0.00 to 9.99, i.e. 10, while the quantization error is constant, ± 0.005; in other words, it is 0.001. In floating point representation the signal range is from 0.0×10^0 to 9.9×10^9, i.e. 10^{10}, while the quantization error varies.

The quantization error is $2 \times 0.05 \times 10^0$ corresponding to a signal level of 0.00×10^0 to 9.9×10^0 and varies up to an error of $2 \times 0.05 \times 10^9$ corresponding to a signal level of 1.0×10^9 to 9.9×10^9. This means that the dynamic range is 10^{10} divided by 0.1, that is, 10^{11}. It is 10^8 times greater than that in the fixed point case.

The instantaneous signal-to-noise ratio is different from the dynamic range. It is defined as the ratio of noise to a common level. Because the quantization noise is constant in the fixed point case, the instantaneous S/N

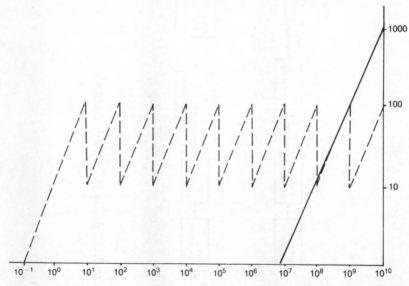

Figure 7.15 Comparison of the dynamic ranges between both floating and fixed point representation.

will decrease with signal level. For example, the S/N on a signal level of 9.99 is $10/0.01 = 1000$, while the S/N for a signal level of 4.99 is $5/0.01 = 500$, as shown in Figure 7.15.

The quantization noise in the floating point representation varies. The S/N at a signal level 9.9×10^9 is $10 \times 10^9/0.1 \times 10^9 = 100$ and the S/N at a signal level of 1.0×10^9 is $1.0 \times 10^9/0.1 \times 10^9 = 10$. The S/N at signal levels of 9.9×10^8 to 1.0×10^8 also fall in the range 100 to 10. The result is that the instaneous S/N in the floating point representation occupies a band from 10 to 100 as shown in Figure 7.15. In other words, the wider dynamic range can be obtained at the cost of the instantaneous S/N using floating point representation.

Quantization is exactly the same expressed in the binary system except that the decimal number represented by base 10^n is converted to the binary number to base 2^n, where n is the number of bits.

Although some distortion and noise occur during the recording and playback of the audio digital signal, as long as they can be detected, the original signal can be restored. This is the key to obtaining high fidelity. To achieve this, error code detection and correction are usually included in the PCM channel. The error code may be a single code, known as a random error, or a block of codes known as a burst error. A parity check is used to correct the random error and an interlace code is used to correct for burst error in the PCM channel of the 8 mm VCR.

Parity check

Before recording a set of PCM data code, 8 bits, added to a parity bit, make a parity encode system. The encode system is called the odd parity system if the number of 1s in 9 bits, including the 8 bits of data code and the parity bit, is odd. If the number of bits is even then it is known as the even parity system. The parity bit itself should be generated from the 8 bits of data code. For example, if the data code is 10 001 010, the parity bit is 0 in the odd parity system, and 1 in the even parity bit system. The even parity bit system is used in the 8 mm VCR.

Figure 7.16 shows how the parity bit is generated. Figure 7.16a shows a parity tree. If the number of 1s in four data codes, a, b, c and d, is even the output from the p-terminal is 0, i.e. $p = 0$. Conversely, if the number of 1s is odd, $p = 1$. Figure 7.16b and c show, respectively, how the odd and even parity systems are formed, which consist of two parity trees and an exclusive OR (XOR) gate. The output of the XOR gate in Figure 7.16c is then inverted.

The data codes of 10 001 010 are still used as an example in Figure 7.16. Because $A = 1$ and $B = 0$, the parity bit, p, should be 1 for Figure 7.16b or 0 for Figure 7.16c.

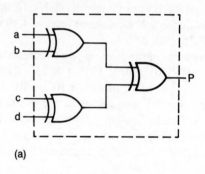

(a)

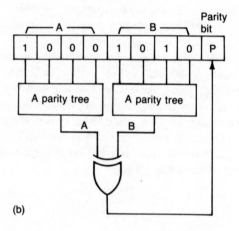

(b)

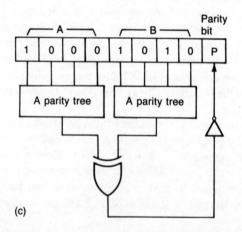

(c)

Figure 7.16 Generation of the parity bit: (a) a parity tree, (b) parity system, (c) the odd parity system.

Table 7.1 Audio PCM frame configuration in the 8 mm VCR

Synchronization code	Add code	Error correct code (Q)	Data code				Error correct code (P)	Data code				CRC code	
			W1	W2	W3	W4		W5	W6	W7	W8	S1	S2
Number of bits													
3	8	8	8	8	8	8	8	8	8	8	8	8	8

Error codes

The frame configuration of the audio PCM signal used in the 8 mm VCR is shown in Table 7.1. This shows the error correcting codes P and Q are generated according to data codes $W1$ to $W8$ respectively. For example, suppose the data codes are those shown in Table 7.2, then P is 10 000 001 in the even parity system.

Table 7.2 Example data codes W1 to W8 (P is in the *even* parity system)

W1	W2	W3	W4	W5	W6	W7	W8	P
1	1	0	0	1	0	1	1	1
0	0	1	1	0	1	0	1	0
1	0	0	1	1	1	0	0	0
1	1	1	0	1	0	1	1	0
0	1	0	1	1	1	1	1	0
1	0	0	0	0	1	1	1	0
1	1	1	1	0	0	0	0	0
1	1	0	0	0	1	0	0	1

The formula relating to P and Q are expressed generally as:

$$P = W1 + W2 + W3 + W4 + W5 + W6 + W7 + W8$$
$$Q = 8W1 + 7W2 + 6W3 + 5W4 + 4W5 + 3W6 + 2W7 + 1W8$$

where the multipliers 8, 7, 6, ..., 2, 1 in the formula for Q are weight coefficients.

The errors of the data codes and correcting codes which occur in the recording and replaying of audio PCM are written as $E1$, $E2$, ..., $E7$, $E8$ and Ep, Eq respectively. No error is indicated as $E = 0$, and the reverse is $E = 1$. The data codes and correcting codes replayed are as shown in Table

Table 7.3 Data and correcting codes replayed

$$W1' = W1 + E1$$
$$W2' = W2 + E2$$
$$W3' = W3 + E3$$
$$W4' = W4 + E4$$
$$W5' = W5 + E5$$
$$W6' = W6 + E6$$
$$W7' = W7 + E7$$
$$W8' = W8 + E8$$

$$P' = P + Ep$$
$$Q' = Q + Eq$$

7.3. The CRC codes in this table, $S1$ and $S2$, consist of the replayed data and correcting codes, where:

$$S1 = W1' + W2' + W3' + \ldots + W8' - P'$$
$$= (W1 + W2 + W3 + \ldots + W8 - P) + (E1 + E2 + E3 + \ldots + E8 - Ep)$$
$$= E1 + E2 + E3 + \ldots + E8 - Ep$$

Similarly,

$$S1 = 8E1 + 7E2 + 6E3 + \ldots + 1E8$$

If the error only occurs in $W1$, i.e. $E1 \neq 0$ and $E2 = E3 = \ldots = E8 = 0$, the CRC codes will be $S1 = E1$ and $S2 = 8E1$ and the result is $S2 = 8S1$. The same conclusion is available for other data codes. These are shown in Table 7.4. Thus the error in one of eight data codes can be determined according to the relationship between $S1$ and $S2$.

Table 7.4 Data codes

$S2 = 7S1$ for W2
$S2 = 6S1$ for W3
$S2 = 6S1$ for W4
$S2 = 5S1$ for W5
$S2 = 4S1$ for W6
$S2 = 3S1$ for W7
$S2 = 2S1$ for W8

Interlace code

The interlace encode system is needed to correct for burst error. Using interlace encoding, a block of error codes can be dispersed to a lot of single codes, and then these can be corrected using the parity check and correction just previously described, i.e. single code.

Suppose that the order of the original code is:

$$\ldots -15 \ -14 \ -13 \ -12 \ -11 \ -10 \ -9 \ -8 \ -7 \ -6 \ -5 \ -4$$
$$-3 \ -2 \ -1 \ 0 \ 1 \ 2 \ 3 \ldots$$

the first step is to realign the original order into three rows:

$$\ldots -18 \ -15 \ -12 \ -9 \ -6 \ -3 \ 0 \ 3 \ 6 \ 9 \ 12 \ 15 \ 18 \ldots$$
$$\ldots -17 \ -14 \ -11 \ -8 \ -5 \ -2 \ 1 \ 4 \ 7 \ 10 \ 13 \ 16 \ 19 \ldots$$
$$\ldots -16 \ -13 \ -10 \ -7 \ -4 \ -1 \ 2 \ 5 \ 8 \ 11 \ 14 \ 17 \ 20 \ldots$$

The second step is to shift the second row to the right for three columns and to shift the third row to the right for six columns:

$$\ldots \ -12 \ -9 \ -6 \ -3 \ \ \ 0 \ \ \ 3 \ \ \ 6 \ \ \ 9 \ \ 12 \ 15 \ 18 \ \ldots$$
$$\ldots \ -17 \ -14 \ -11 \ -8 \ \ -5 \ -2 \ \ 1 \ \ 4 \ \ 7 \ 10 \ 13 \ \ldots$$
$$\ldots \ -19 \ -16 \ -13 \ -10 \ -7 \ -4 \ -1 \ \ 2 \ \ 5 \ \ 8 \ 11 \ \ldots$$

The third step involves realigning the order according to the new order of every column:

$$\ldots \ 0 \ -8 \ -16 \ 3 \ -5 \ -13 \ 6 \ -2 \ -10 \ 9 \ 1 \ -7 \ 12 \ 4 \ -4 \ 15 \ 7 \ -1$$
$$18 \ 10 \ 2 \ \ldots$$

This order of data codes will be recorded and a block of error codes is supposed to have occurred in the replayed codes. The error block will be dispersed to a lot of single codes after the replayed codes are restored into the original code order.

$$\ldots \ -13 \ -12 \ -11 \ -10 \ -9 \ -8 \ -7 \ -6 \ -5 \ -4 \ -3 \ -2 \ -1$$
$$0 \ 1 \ 2 \ 3 \ 4 \ 5 \ 6 \ 7 \ 8 \ 9$$
$$\ldots$$

(the error codes)

In the 8 mm VCR the interlace code for eight data codes is also executed before they are recorded and the encode process is as follows. Using $W1$, $W2$, ..., $W8$, shown in Table 7.2 the encode process is that the first column is not shifted, i.e. 11 001 011; the second column is shifted down one bit, i.e. changed from 00 110 101 to 00 100 111; the third column is shifted down two bits, with each subsequent column being shifted down one further bit until the eighth column is shifted down seven bits, i.e. changed from 11 000 100 to 10 001 001. The shifted data is that shown in Table 7.5. This data is realigned according to the order of $W1'$ to $W8'$. During replay the encoded order is restored and the burst error will be dispersed.

Table 7.5 Shifted replay data

W1′	W2′	W3′	W4′	W5′	W6′	W7′	W8′
1	1	0	0	1	0	1	1
1	0	0	1	1	0	1	0
0	0	1	0	0	1	1	1
1	1	1	0	1	0	1	1
1	1	1	1	0	1	0	1
0	0	1	1	1	1	0	0
1	1	0	0	0	0	1	1
1	0	0	0	1	0	0	1

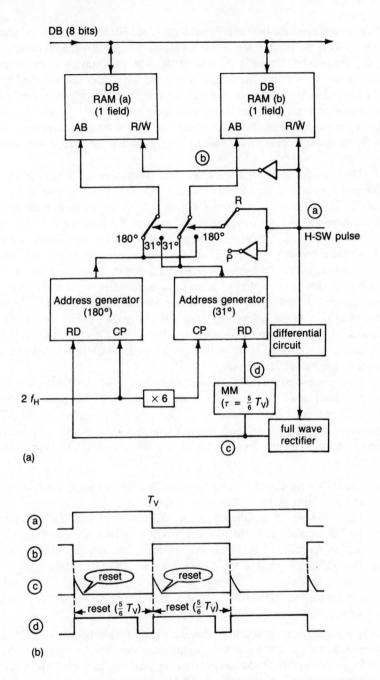

Figure 7.17 Time-axis compression and expansion: (a) principle, (b) waveforms.

In order to compress the time axis from an angle of 180° to 31°, and to expand the time axis in the reverse process, it is necessary to use RAM with memory capacity for two fields, one in the read mode, while the other is in the write mode for one field time. The read-in and write-out modes are switched by the H-SW pulse. Both RAM (1 field) are written in at a sampling rate of $2f_H$ and read out at a rate six times higher than the writing speed. Therefore, two address generators are needed to carry this out. One of the address generators uses the sampling feqeuncy of $2f_H$ as the clock pulse and the other is triggered by $6 \times 2f_H$.

The first address generator is reset at the rising edge or the trailing edge of the H-SW pulse and the second is reset for the period of the reset to $\frac{5}{6}T_V$ (where T_V is 1 field time, i.e. 20 ms). Figure 7.17 shows the method of time-axis compression and the waveforms for resetting the memory.

This process can be described by example. Suppose that RAM(a) and RAM(b) are in the write-in and read-out modes, respectively, during the time of one field, i.e. the R/W terminals of RAM(a) and (b) have a low and high level of the H-SW pulse applied to the respective terminals. 8 bits of data flow, converted from a field signal will be written continuously into RAM(a) with the starting time at the rising edge of the H-SW pulse, and at the address generated by the sampling frequency, $2f_H$. The address is applied to the AB (address bus) terminal of RAM(a) through a switch which is selected to the 180° terminal by the H-SW pulse.

Meanwhile, 8 bits of data flow are read out from RAM(b) with the starting time at an interval of $\frac{5}{6}T_V$ from the rising edge of the H-SW pulse and the address is generated by six times $2f_H$. This address is applied to the AB terminal of RAM(b) via a switch connected to the 31° terminal. The same explanation holds for the reverse case during the next field. The variations are as follows:

1. The R/W terminals of both RAM(a) and RAM(b) are applied with a high and low level of the H-SW pulse respectively.
2. The next 8 bits of data flow from the A/D converter will be written into the DB terminal of RAM(b) and 8 bits of data flow are read out from the DB terminal of RAM(a), and are fed to the head for recording.
3. The address applied to the AB terminal of RAM(a) comes from the address generator (31°) through the switch connecting the 31° terminal, and the address sent to RAM(b) comes from the address generator (180°) through the switch connecting the 180° terminal.

It is not difficult to see from this discussion that although both RAM are simultaneously instructed to write-in then read-out data for feeding to the heads for recording, the RAM contents are the data read in from the previous field. Thus the data is delayed by one field period.

For the same reason, the data replayed and expanded in the time-axis are also delayed for one field. So the PCM audio has to be delayed for one frame after passing through the time-axis compressor in the REC channel and in the time-axis expander in the PB channel. The operating principle of the time-axis expander is the same as that of the compressor. The only difference is that the writing speed is six times higher than the reading speed, so the H-SW pulse has to be reversed before being applied to the two address switches, as shown in Figure 7.17a.

7.4 The auto tracking finder (ATF) circuits

The tracking scheme using CTL has been replaced by the ATF circuit in the 8 mm VCR. Therefore the CTL track as found on the traditional VCR can be omitted on this system. The auto tracking finder (ATF) circuit, or the dynamic tracking finder (DTF) circuit in the PB channel, is shown in Figure 7.18.

During recording four ATF pilots, f_1, f_2, f_3 and f_4 are generated sequentially per field inside the machine, and alternately added into the two video heads, CH-A and CH-B. That is, the ATF pilot is mixed with the video, AFM or APCM signals, and then recorded on the tape. This is shown in Figure 7.19.

The frequencies of the four pilots are:

$f_1 = (378/58) f = 102.54$ kHz $= 102.5$ kHz recorded on the 1st, 5th, ... tracks

$f_2 = (378/50) f = 118.95$ kHz $= 119$ kHz recorded on the 2nd, 6th, ... tracks

$f_3 = (378/36) f = 165.21$ kHz $= 165$ kHz recorded on the 3rd, 7th, ... tracks

$f_4 = (378/40) f = 148.69$ kHz $= 148.5$ kHz recorded on the 4th, 8th, ... tracks

As the four pilot frequencies are lower than the frequency of the luminance and chrominance carriers, and the AFM signal can be easily filtered out by an LPF, it can be seen from Figure 7.19b that the differences between adjacent tracks are close to two frequencies, 46 and 16.5 kHz. These may be represented as follows:

$f_A = f_2 - f_1 = f_3 - f_4 = 16.5$ kHz
$f_B = f_4 - f_1 = f_3 - f_2 = 46$ kHz

During replay the tracking can be detected by comparing the level of the pilots leaked from the left and right sides adjacent to the tracks. This is shown in Figure 7.18.

The PB pilot is separated by an LPF from the PB RF signal and then sent to a balanced modulator, in which the frequency difference between the

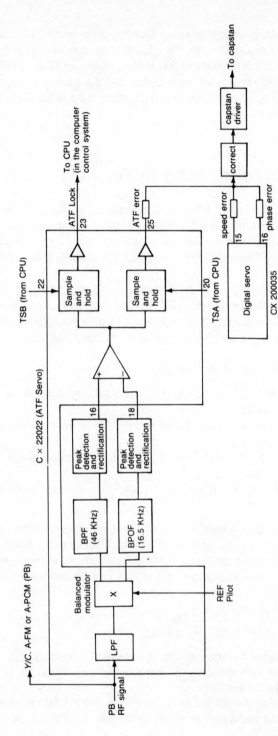

Figure 7.18 ATF circuit in the PB channel.

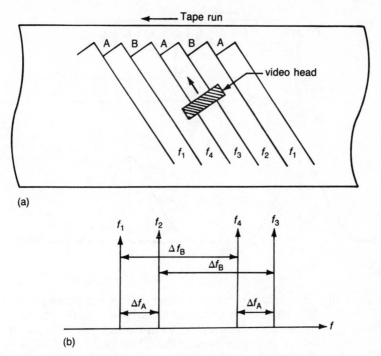

(a)

(b)

Figure 7.19   Four pilots in the ATF system: (a) four pilots located on four tracks, (b) the frequency relationship of four pilots ($\Delta f_A = f_2-f_1 = f_3-f_4 \simeq 16.5$ kHz; $\Delta f_B = f_4-f_1 = f_3-f_2 \simeq 46$ kHz).

REF pilot and PB pilot leaked from the left and right adjacent tracks will be generated. Here, four REF pilots are also generated inside the machine, but their order is the reverse of that in the REC mode, i.e. f_4, f_3, f_2 and f_1. For example, see Figure 7.20, when the PB pilot is f_4 the REF pilot should be f_2, so the frequency difference between the REF pilot and PB main pilot is $f_4-f_2 = 29.5$ kHz, and with the PB pilots on the left and right sides adjacent tracks are, respectively, $f_3-f_2 = 46$ kHz, and $f_2-f_1 = 16.5$ kHz. The latter two frequencies are selected by means of two BPF and then converted to d.c. levels after they are each passed through their respective detectors and rectifiers. This is shown in Figure 7.18.

If the scanning head deviates to the left of the main track, the level corresponding to the frequency difference of 46 kHz is larger than that of 16.5 kHz. A positive voltage will be output from an operational differential amplifier and then used as an ATF error voltage. The error voltage is used to drive the capstan faster through a sampling and hold circuit. In the reverse case, a negative error voltage output from the operational amplifier will make the capstan rotate more slowly when the scanning head is to the right of the

Standard and high band 8 mm VCR   **247**

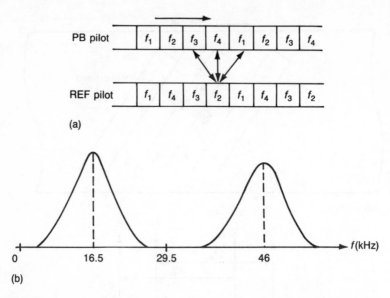

Figure 7.20 An example of the frequency differences between the REF pilot and the pilots of the adjacent tracks: (a) the frequency difference between the REF pilot and the PB pilot of the main track or the adjacent tracks ($f_1 - f_2 = 29.5$ kHz, $f_3 - f_2 = 46$ kHz, $f_2 - f_1 = 16.5$ kHz), (b) the two tuning ranges of the BPFs to select the REF pilot and PB pilot from adjacent tracks.

main track, and the level corresponding to the frequency difference of 16.5 kHz is larger than that of 46 kHz. It can also be seen from Figure 7.20 that the difference between the REF pilot and the PB main track pilot is always 29.5 kHz or 0 Hz. This frequency is outside the range of the two BPFs and cannot pass through them.

Using these principles, the scanning video heads can always follow the main video track on the tape. If the PB track deviates from the REF pilot, as shown in Figure 7.21b, both left and right sides of the adjacent tracks will be exchanged with each other resulting in reverse correction. The deviated track needs to be detected and corrected before the ATF error is acted upon. This is called ATF lock.

In the ATF lock, the REF pilots are prolonged by 3 ms and overlapped on part of the next track. Another sampling and hold circuit is added and controlled by a TSB pulse sent from the CPU on the computer control system, as shown in Figure 7.18. The PB main pilot is compared with the REF pilot for the prolonged period, while the TSB pulse is at a low level.

During normal playback the frequency difference between the prolonged PB and REF pilots is always 16.5 kHz, corresponding to a low level output

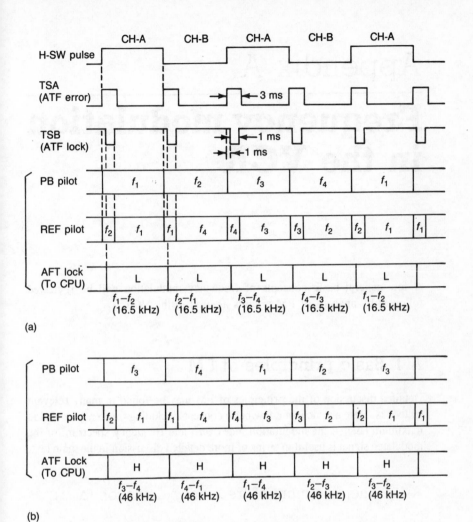

Figure 7.21 Normal PB pilot and a track deviation: (a) normal playback, (b) track deviated.

from the ATF lock terminal in Figure 7.18. If the REF pilot generated inside the machine deviates, the frequency difference changes to 46 kHz and corresponds to a high level output to the CPU, followed by speeding up of the capstan slightly until normal playback reappears.

Appendix A

Frequency modulation in the VCR

A few general basic concepts of frequency modulation will be reviewed followed by analysis of FM circuits as used in the VCR.

A.1 Basic principles of FM

Detailed discussion of the principles of FM may be found in many relevant textbooks. Here a review of a few basic concepts including frequency deviation maximum ($\Delta\omega_m$), the modulation index (m_f) and frequency spectrum of the modulated signal is useful in terms of more detailed discussion in the main text.

A.1.1 The maximum of frequency deviation ($\Delta\omega_m$)

Supposing the modulating signal is in the form:

$$B(t) = B_0 \cos \Omega t \qquad \text{(A.1)}$$

where Ω is the angular frequency of the modulating signal and the carrier is in the form:

$$A(t) = A_0 \cos \omega_0 t \qquad \text{(A.2)}$$

where ω_0 is the angular frequency of the carrier, then the angular frequency after FM is:

$$\omega(t) = \omega_0 + KB_0 \cos \Omega t \qquad \text{(A.3)}$$

where K is the modulation sensitivity and indicates how much frequency deviation is caused by a unit modulating voltage. The units are Hz/volt (arc/s volt).

The angular frequency of the modulated signal varies around a centre frequency, ω_0, in response to the modulating signal. So in Equation A.3 the second term is the amount of frequency deviation in the modulated signal. This is $\Delta\omega$, where

$$\Delta\omega = KB_0 \cos \Omega t \qquad (A.4)$$

When $\cos \Omega t = 1$ the frequency deviation is a maximum, i.e.

$$\Delta\omega_m = KB_0 \qquad (A.5)$$

Conclusion 1 the maximum frequency deviation of the modulated signal ($\Delta\omega_m$) is directly proportional to the amplitude of the modulating signal (B_0). This explains why the gain adjust of the AGC amplifier is known as the frequency deviation adjust.

A.1.2 The modulating index (m_f)

The amplitude of the modulating signal corresponds to the maximum frequency deviation of the modulated signal. We need to consider what the angular frequency of the modulating signal corresponds to. From Equation A.3 we can obtain the phase angle of the modulated signal:

$$
\begin{aligned}
\theta(t) &= \int \omega(t)\mathrm{d}t \\
&= \int (\omega_0 + \Delta\omega_m \cos \Omega t)\mathrm{d}t \\
&= \omega_0 t + \frac{\Delta\omega_m}{\Omega} \sin \Omega t + \theta_0 \\
&= \omega_0 t + m_f \sin \Omega t + \theta_0 \qquad (A.6)
\end{aligned}
$$

where m_f is the modulation index of the frequency modulation. The term $\omega_0 t$ is the phase angle of the carrier and $m_f \sin \Omega t$ is the phase shift caused by the modulating signal.

Conclusion 2 After limiting the maximum frequency deviation, the modulation index (m_f) is inversely proportional to the angular frequency (ω) of the modulating signal. This explains why the S/N ratio in the high frequency region of the modulating signal deteriorates during frequency modulation

A.1.3 The frequency spectrum of the frequency modulated signal

By substituting Equation A.6 into A.2 we can obtain the frequency spectrum of the frequency modulated signal:

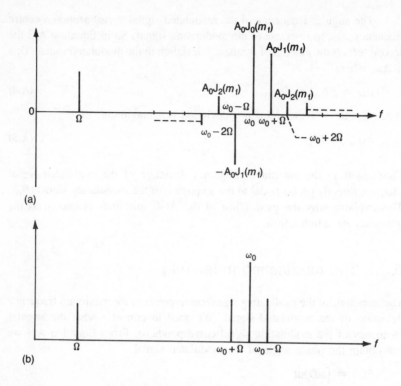

Figure A.1 Frequency spectrum of the FM signal: (a) FM, (b) AM.

$$A(t) = A_0 \cos \theta(t)$$

$$= A_0 \cos (\omega_0 t + m_f \sin \Omega t + \theta_0)$$

$$= A_0 [\cos \omega_0 t \cos(m_f \sin \Omega t + \theta_0) - \sin \omega_0 t \sin(m_f \sin \Omega t + \theta_0)]$$

$$= A_0 \{J_0(m_f) \cos \omega_0 t + [J_1(m_f)\cos(\omega_0 + \Omega)t - J_1(m_f)\cos(\omega_0 - \Omega)t$$

$$+ [J_2(m_f)\cos(\omega_0 + 2\Omega)t - J_2(m_f)\cos(\omega_0 - 2\Omega)t]$$

$$+ \sum^{n}[J_n(m_f)\cos(\omega_0 + n\Omega)t - J_n(m_f)\cos(\omega_0 - n\Omega)t]\} \qquad \text{(A.7)}$$

where J_n is the nth component of a Bessel function. This equation is shown graphically in Figure A.1 and shows the frequency spectrum of the FM signal. The frequency spectrum of the equivalent AM signal is also shown for comparison.

In theory there are an infinite number of upper and lower side frequencies around the carrier frequency (ω_0) in the modulated signal. The amplitudes of the carrier frequency, first harmonic and second harmonic, etc., are dependent on the magnitude of the 0th, 1st and 2nd, etc., Bessel function component, according to tbe Bessel function $J_0(m_f)$, $J_1(m_f)$ and $J_2(m_f)$, etc. So if we

252 Video recorders

choose a modulation index where $m_f < 1$, the modulated signal has one pair of side frequencies only:

$$J_1(m_f)\cos(\omega_0+\Omega)t$$

and

$$J_1(m_f)\cos(\omega_0-\Omega)t$$

Then:

$$A(t)-A_0\{J_0(m_f)\cos\ \omega_0 t+[J_1(m_f)\cos(\omega_0+\Omega)t-J_1(m_f)\cos(\omega_0-\Omega)t]\}$$

(A.8)

This yields a spectrum which is not the same as an amplitude modulated signal. (Note: the lower side frequency has a negative sign indicating an opposite phase to the upper side frequency.)

A.2 Frequency modulator circuit analysis in VCR

A voltage crystal-controlled oscillator (VXO) is generally used as a kind of slope modulator to generate frequency modulated video. We have seen, however, that since A is large, the utilizable range of linear slope is too narrow. In order to get round this problem, the push—pull type of VXO, which consists of a double VXO, is used in a few high quality VTRs. Most VCRs have utilized a monostable multivibrator as a frequency modulator. Therefore, we must analyse the principle of using a multivibrator as a frequency modulator.

A.2.1 A simple review of the principles of the multivibrator

Detailed principles of the operation of the multivibrator are given in many general electronic textbooks, so we will look only at the main points of operation. Our main interest lies in which parts of the multivibrator change the frequency generated and, therefore, how it may be used as a frequency modulator.

Figure A.2 shows the circuit of a multivibrator. In operation, the main features of the multivibrator are:

1.   There are two steady states: Q1 on and Q2 off (and vice versa).
2.   The factor which changes the steady states is the charging or discharging of capacitances C1 or C2. When C1 is charging, Q2 is turned off and when C1 is discharging Q2 is turned on. This is shown in Figure A.2. Q1 is similarly turned on and off by C2.

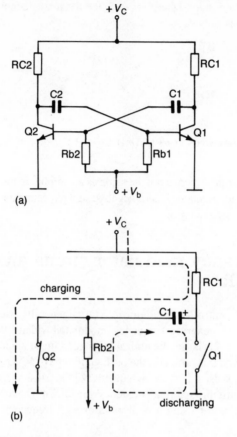

Figure A.2   The multivibrator circuit.

3.   The conditions causing oscillation are states Q1 and Q2, and the feedback circuits. This process is very quick and the frequency (or period) of oscillation is mainly dependent on the charging and discharging times of the capacitances C1 or C2. Because the circuit of the multivibrator is symmetrical, we can discuss the charging and discharging of C1 only, without loss of generalization.

A.2.2 The period of the multivibrator

Firstly we can begin by considering C1 charged and then beginning to discharge. As is shown in many electronic textbooks, the voltage at the base of Q2 changes during the discharge of C1 as follows:

$$V_{b2}(t) = V_{b2}(\infty) - \{V_{b2}(\infty) - V_{b2}(O)\}e^{-t/R_{b2}C1} \qquad \textbf{(A.9)}$$

In Figure A.2, at $t = 0$, C1 is fully charged and

$$V_{b2}(t)_{t=0} = -V_c$$

and at $t = $ C1 is fully discharged and

$$V_{b2}(t)_{t=\infty} = V_b$$

Substituting these values into Equation A.9 yields:

$$V_{b2}(t) = V_b - \{V_b + V_c\}e^{-t/\tau} \qquad (A.10)$$

where $\tau = R_{b2}C1$ and is called the discharge time constant.

After discharging of C1, the transistors Q2 and Q1 turn on and off respectively. C1 then begins to recharge. Because the components of the multivibrator (including Q1 and Q2) are symmetrical, the time of discharging C1 corresponds to half of the oscillation period (T) of the multivibrator. The base voltage (V_b) is maintained at zero volts during Q2 turn on. During this period Equation A.10 will become

$$0 = V_b - (V_b + V_c)e^{-T/2\tau} \qquad (A.11)$$

where $T = 1/f$ is the period of oscillation of the multivibrator (and f is the oscillation frequency). From Equation A.11 we can obtain:

$$T = 2\tau \ln\left(1 + \frac{V_c}{V_b}\right) \approx 2\tau\left[\left(1 + \frac{V_c}{V_b}\right) - 1\right]\Big/\left(1 + \frac{V_c}{V_b}\right)$$

$$= 2\tau \frac{V_c}{(V_c + V_b)}$$

or

$$f = \frac{1}{2\tau}\left(1 + \frac{V_b}{V_c}\right) = A + BV_b \qquad (A.12)$$

where

$$A = \tfrac{1}{2}\frac{1}{R_{b2}C1} \qquad (A.13)$$

and

$$B = \tfrac{1}{2}\frac{1}{R_{b2}C1}\frac{1}{V_c} \qquad (A.14)$$

In this analysis of the period T we have used the natural logarithmic characteristic:

$$\ln x = \frac{x-1}{x} + \tfrac{1}{2}\left(\frac{x-1}{x}\right)^2 + \tfrac{1}{3}\left(\frac{x-1}{x}\right)^3 + \dots$$

when $x \geq \tfrac{1}{2}$ the square and higher terms are neglected, i.e.

$$\ln x \simeq \frac{x-1}{x}$$

Because all powers of V_c and V_b are positive:

$$\left(1 + \frac{V_c}{V_b}\right) > \tfrac{1}{2}$$

From Equation A.12 we can obtain the following conclusions:

1. In theory all factors, R_b, C, V_c and V_b could influence the frequency of oscillation. Using V_b only is the best and gives higher sensitivity and a wider linear region.
2. The variation of the oscillation frequency, f, with offset voltage (V_b) is approximately linear as shown in Figure A.3. The slope of the function

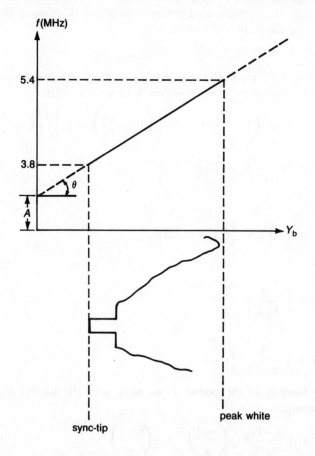

Figure A.3   Variation of oscillator frequency with offset voltage.

is adjusted by the voltage V_c after the centre frequency of oscillation is decided, by fixing the values of R_b and C.

$$\tan \theta = \frac{1}{2} \frac{1}{R_{b2}C1} \frac{1}{V_c} \tag{A.15}$$

Therefore, we could ensure that the frequency of oscillation corresponds to the modulated video signal by adjusting the frequency deviation (at AGC loop), clamping voltage (at sync-tip clamp) and the supply voltage of the multivibrator; adjusting the clamping voltage of the video signal and the slope of the modulating characteristic to be positioned at the sync-tip level; and adjusting the frequency deviation to ensure that the modulating frequency corresponds to that of the video signal.

Appendix B

Basic concepts of a differential amplifier

B.1 The current flow

As shown in Figure B.1, the current through the transistor T3, (I_3) is produced by converging currents, I_{e1} and I_{e2}, which flow through T1 and T2.

$$I_3 = I_{e1} + I_{e2} \tag{B.1}$$

When input voltage V_2 is unchanged, I_3 is constant and I_{e1} and I_{e2} change as the input voltage V_1 changes, i.e.

$$I_{e1} = I_{e1(0)} \tag{B.2}$$

and

$$I_{e2} = I_{e2(0)} \tag{B.3}$$

where $I_{e1(0)}$ and $I_{e2(0)}$ are the quiescent currents in T1 and T2 when $V_1 = 0$. Since T1 and T2 are symmetrical, $I_{e1(0)} = I_{e2(0)} = 0.513$.

B.2 Relationship between voltage and current

The relationship between the base-emitter voltage (V_{be}) and the emitter current (I_e) of the transistor satisfies the general diode equation:

$$I_e = I_s \left[e^{(V_{be}/KTq)} - 1 \right] \tag{B.4}$$

where I_s = inverse saturation current in an $e-b$ junction;
q = electronic charge, $q = 1.6 \times 10^{-19}$ Coulombs;
K = Kelvin's constant, $K = 1.38 \times 10^{-23}$ Joules/K; and
T = absolute temperature in K. At normal room temperature, for

example 25°C (or 298.16 K)

$$\frac{KT}{q} = 26 \text{ mV}$$

and

$$e^{(V/(K\,T/q))} \gg 1$$

then Equation B.4 becomes:

$$I_e \simeq I_s e^{(V_{be}/26)}$$

or

$$I_{be} = 26 \ln \frac{I_e}{I_s} \qquad\qquad\qquad \textbf{(B.5)}$$

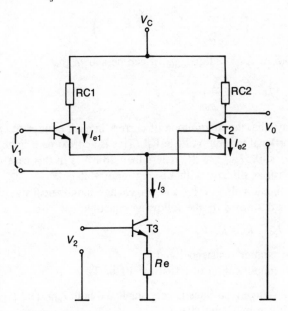

Figure B.1   Differential amplifier.

As shown in Figure B.1:

$$V_1 = V_{be1} - V_{be2}$$

$$= 26 \ln \frac{I_{e1}}{I_s} - 26 \ln \frac{I_{e2}}{I_s}$$

$$= 26(\ln I_{e1} - \ln I_{e2})$$

Using Equations B.2 and B.3 we can show that (assuming an increase in current in T1 and a decrease in current in T2):

$$V_1 = 26[\ln(I_{e1(0)} + \Delta I_e) - \ln(I_{e2(0)} + \Delta I_e)]$$
$$= 26[\ln(1+x) - \ln(1-x)]$$

if

$$x = \frac{\Delta I_e}{I_{e1(0)}} = \frac{\Delta I_e}{I_{e2(0)}}$$

where

$$\ln(1+x) = \left(x - \frac{x^2}{2} + \frac{x^3}{3} - \frac{x^4}{4} + \dots\right)$$

and

$$\ln(1-x) = \left(x + \frac{x^2}{2} + \frac{x^3}{3} + \frac{x^4}{4} + \dots\right)$$

when $x < 0.5$ or $I_e I_e < 0.5\, I_{e2(0)}$, it can be shown that:

$$V_1 \simeq 52x = 52\,\frac{\Delta I_e}{I_{e2(0)}} \tag{B.6}$$

That is to say, that the distribution of current in each emitter of Q1 and Q2 (I_e) is proportional to input voltage V_1. (This is approximate since I_e is very small, i.e. $I_e < 0.5 I_{e2(0)}$, as V_1 is less than 26 mV.) In this case the output voltage V_0, varies directly with varying input voltage V_1.

When input voltages V_2 and V_1 change simultaneously, the output voltage V_0 is described by the following equation:

$$V_0 = I_{e2(0)}\, R_e - K_v\, V_1 \tag{B.7}$$

where R_e = emitter resistance; and
K_v = amplification of transistor T1 or T2.

As the signal source resistance is very small and both $I_{e1(0)}$ and $I_{e2(0)}$ are less than 1 mA it can be shown that:

$$K_v \simeq \frac{R_e}{52}\, I_{e2(0)} \ (mA/mV) \tag{B.8}$$

Substituting for K_v from Equation B.8 into Equation B.7 and noting that

$$I_{e2(0)} = 0.513 = \tfrac{1}{2}\,\frac{V_2}{R_e}$$

then we can obtain:

$$V_0 = I_{e2(0)}R_e + \frac{R_e}{52} I_{e2(0)}V_1$$

$$= \frac{V_2}{2} + \frac{V_1V_2}{104} \tag{B.9}$$

or using the unit volts instead of millivolts ($104 \text{ mV} \simeq 0.1 \text{ V}$) and using V_{02} to indicate the output voltage from the collector of T2, Equation B.9 can be written as:

$$V_{02} = \frac{V_2}{2} + 10V_1V_2 \tag{B.10}$$

Similarly, we could write the output voltage (V_{01}) from the collector of T1 as:

$$V_{01} = I_{e1(0)}R_e + K_vV_1$$

$$= I_{e1(0)}R_e - \frac{R_eI_{e1(0)}}{52} V_1$$

$$= \frac{V_2}{2} - 10V_1V_2 \tag{B.11}$$

From these expressions, the output voltages of the differential amplifier has the following features:

1. The second term includes the *multiplication* of two input signals, V_1 and V_2.
2. In the first term there are no other frequency components except V_2.

These features would be used in a modulator, discriminator, mixer and auto-gain controller, etc.

B.3 The differential amplifier as a modulator

Supposing $V_1 = A \cos \omega t$ is used as a carrier signal and $V_2 = B_0 + B \cos \Omega t$ is used as a modulating signal (B_0 is the d.c. offset voltage of emitter coupling of transistor T3 in Figure B.1), then we can substitute for V_1 and V_2 in Equation B.10 and show that:

$$V_{02} = \tfrac{1}{2}B_0 + \tfrac{1}{2}B \cos \Omega t + 10AB_0 \cos \omega t + 10AB \cos \omega t \cos \Omega t$$

$$= \tfrac{1}{2}B_0 + \tfrac{1}{2}B \cos \Omega t + [10AB_0 \cos \omega t + 5AB \cos(\omega + \Omega)t$$

$$+ 5AB \cos(\omega - \Omega)t] \tag{B.12}$$

Substituting into Equation B.11, we also obtain:

$$V_{01} = \tfrac{1}{2}B_0 + \tfrac{1}{2}B \cos \Omega t - [10AB_0 \cos \omega t + 5AB \cos(\omega+\Omega)t$$
$$+ 5AB \cos(\omega-\Omega)t] \tag{B.13}$$

Using a filter to separate the last three terms in the bracket, we can obtain a modulated signal which contains the main carrier, ω, and two side frequencies, $(\omega+\Omega)$ and $(\omega-\Omega)$..

We can also change the form of Equations B.12 and B.13 to include current,

$$I_{e1} = \frac{V_{01}}{R_e}$$

and

$$I_{e2} = \frac{V_{02}}{R_e}$$

So:

$$I_{e1} = \frac{1}{R_e}\left[\frac{B_0}{2} + \frac{B}{2}\cos \Omega t - 10AB_0 \cos \omega t - 5AB \cos(\omega+\Omega)t\right.$$
$$\left. - 5AB \cos(\omega-\Omega)t\right]$$
$$= I_0 + I_\Omega - I_\omega - I_{\omega+\Omega} = I_{\omega-\Omega} \tag{B.14}$$

and

$$I_{e2} = I_0 + I_\Omega + I_\omega + I_{\omega+\Omega} + I_{\omega-\Omega} \tag{B.15}$$

where

$$I_0 = \tfrac{1}{2}\frac{B_0}{R_e}$$

$$I_\omega = \tfrac{1}{2}\frac{B}{R_e}\cos \Omega t$$

$$I_\omega = 10\frac{AB_0}{R_e}\cos \omega t$$

$$I_{\omega+\Omega} = 5\frac{AB}{R_e}\cos(\omega+\Omega)t$$

$$I_{\omega-\Omega} = 5\frac{AB}{R_e}\cos(\omega-\Omega)t \tag{B.16}$$

Index